I WILL ALWAYS PLACE THE

MISSION

I WILL

NEVER ACCEPT DEFEAT.

LEADERSHIP IN A COMPLEX WORLD

I WILL NEVER QUIT.

MSL 402
REVISED EDITION

I WILL NEVER LEAVE

A FALLEN COMRADE.

Custom Publishing

New York   Boston   San Francisco

London   Toronto   Sydney   Tokyo   Singapore   Madrid

Mexico City   Munich   Paris   Cape Town   Hong Kong   Montreal

**Pearson
Custom Publishing**
is a division of

www.pearsonhighered.com

ISBN 10: 0-536-56322-5
ISBN 13: 978-0-536-56322-4

CONTENTS

Officership Track

INTRODUCTION

Key Points

Leadership is intangible, and therefore no weapon ever designed can replace it.

General of the Army Omar N. Bradley

Overview of the BOLC I: ROTC Curriculum

Being an officer in the US Army means primarily being a leader, as well as a counselor, a strategist, and a motivator. Officers must lead other Soldiers in all situations and adjust to environments that are always changing. To prepare prospective officers to meet this challenge, the Army ROTC program is designed to develop confident, competent, and adaptive leaders with the basic military science and leadership foundation necessary not only to lead small units in the Contemporary Operating Environment (COE) but also to evolve into the Army's future senior leaders.

The ROTC program is the first, or pre-commissioning, phase of the Army's Basic Officer Leader Course (BOLC). The goal of BOLC is to develop competent and confident leaders imbued with a warrior ethos, grounded in fieldcraft, and skilled in leading Soldiers, training subordinates, and employing and maintaining equipment. BOLC is designed to ensure a tough, standardized, small-unit leadership experience that flows progressively from the pre-commissioning phase (BOLC I, one source of which is ROTC) through the initial-entry field leadership phase (BOLC II) to the branch technical phase (BOLC III). This progressive sequence will produce officers with maturity, confidence, and competence who share a common bond—regardless of whether their branch is combat arms, combat support, or combat service support—and who are prepared to lead small units upon arrival at their first unit of assignment.

The basis of the Army ROTC curriculum is the BOLC common core task list, which represents the foundation of competencies a second lieutenant needs upon arrival at his or her first unit. ROTC Cadets receive education and training in BOLC I common core tasks, as do officers produced by other commissioning sources (the United States Military Academy and Officer Candidate School). Then, in BOLC II and III, all second lieutenants, regardless of commissioning source, participate in more advanced, field- and branch-oriented education and training events that are part of the BOLC II and III common core task lists.

Like the BOLC model, ROTC's Military Science and Leadership (MSL) courses are sequential and progressive; that is, the content and expectations placed on you as the student increase as you progress through the ROTC Program. As you may recall, the academic rigor of your MS II year was greater than that of your MS I year. Also, your MS III year was without a doubt far more challenging and involved than your MS I and II years combined. Now you are an MS IV Cadet and your MS IV year will be much more academically rigorous than before. The key difference between previous MS years and your MS IV year is that your professor of military science (PMS) and cadre place the onus of learning on you—the future Army officer who must now learn to take control of his or her own professional development and career management.

In previous years, your MSL instructors were right there with you, along with your fellow Cadets learning together as you proceeded through the academic semesters. In many cases, you completed work or tasks as a group, more than as an individual. As an MS IV Cadet, your PMS and the ROTC cadre will also be there to mentor your professional development, but their expectations of how much effort you put into your own professional and personal development is exponentially greater than when you were an underclass Cadet.

The duties and responsibilities you will be given as an MS IV will require you to work more independently, but you will also be required to come together as Cadet staff to assess, plan, coordinate, and execute your PMS's short- and near-term training on campus. As will be the case when you arrive at your first unit of assignment as a new lieutenant, you will be an individual responsible for yourself and your own work—your company commander will be there to mentor you, but you are not going to be led by the hand and be spoon-fed. As an MS IV, you must get in the habit of working independently with little guidance to produce high-quality work that is correct the first time.

As was the case with your underclass ROTC curriculum, your MS IV courses are organized into five tracks: the Leadership, the Personal Development, the Values and Ethics, the Officership, and the Tactics and Techniques Tracks. As an MS IV, you will play a key role in recruiting, retaining, and mentoring Cadets enrolled in the Basic Course and your PMS may also assign you mentorship and evaluation roles for the battalion's MS III Cadets. You will be expected to pass along to the MS III Cadets your knowledge of LDAC and how to be a success at Warrior Forge.

In addition to classroom instruction, your MS IV year will provide you with multiple opportunities to apply military science and leadership concepts in field environments, including leadership labs, battalion or joint field training exercises (FTX), and any battalion STX training that your PMS may direct. As a contracted Cadet, you must participate in physical training (PT) to build your fitness ethos and maintain Army Physical Fitness Test (APFT) standards. It is important for the MS IV Cadets to lead by example during battalion PT events, but also in non-ROTC social environments to help instill into the underclassman Cadets the Army Values and Warrior Ethos.

Military Science and Leadership Tracks

Each of the five learning tracks in the Army ROTC Military Science and Leadership curriculum has subcategories that are reiterated and developed progressively through the MSL courses. The US Army has long recognized the importance of the effective leader who fully embodies the leadership ethos, who is fully committed to being a lifelong learner of leadership as a process and journey rather than a destination; a person who has the professional acumen to put this leadership into action in an effective, value-added manner regardless of the challenge of the situation faced in the fast-paced, ever-changing COE.

Leadership

- *Leader Attributes* from FM 6-22 are used throughout the curriculum as a graphic organizer for developing a basic knowledge of leader dimensions. The implicit focus throughout the curriculum is on the importance of personal discipline in becoming a leader of character, a leader with presence, and a leader with intellectual capacity.

- *Core Leader Competencies* are centered around what an Army leader does. These competencies are defined and illustrated as they apply to direct (tactical), organizational (operational), and strategic levels of leader responsibility. The course of study as a whole is designed to challenge and develop the leader's ability to lead (demonstrate competence, communicate, and motivate), develop self and others (adapt, learn, and mentor), and achieve (prioritize, plan, and execute).

Personal Development

- *Character* development is an implicit aspect of the ROTC curriculum. Cadets are challenged throughout the course of study to recognize and model the Army Values of loyalty, duty, respect, selfless service, honor, integrity, and personal courage; to empathize with their peers, subordinates, and others; and to live the Warrior Ethos.

- *Physical Presence* is foundational for Army leader development. Every Cadet who seeks to become an officer must be able to demonstrate an exceptional level of physical fitness, composure, confidence, and resilience.

- *Intellectual Capacity* has always been and continues to be an imperative characteristic for officers serving in the US Army. Those serving in the contemporary operating environments of Iraq and Afghanistan are learning firsthand the value of mental agility and innovation to Army leadership. Vignettes and case studies from these environments are used to challenge Cadets to examine nonlinear situations, to hone their judgment, and to increase their tactical, technical, cultural, and geopolitical knowledge.

Values and Ethics

- *Army Values.* While it is important for Cadets to be able to articulate the seven Army Values, it is even more imperative that they be able to demonstrate these values in their daily interactions with others. Values form the foundation for Army leadership.

- *Professional Ethics.* In addition to the Army Values, military codes and regulations govern ethical behavior and decision making. Cadets apply the ethical decision making process during case studies and historical vignettes.

- *Warrior Ethos* is embedded in case studies and historical vignettes throughout the curriculum. Cadre members discuss the four basic principles of the Warrior Ethos whenever possible. Cadets apply the Warrior Ethos to increasingly complex situations as they progress through the ROTC program.

Officership

- *Military Heritage.* Cadre members teach and model military heritage through daily performance and contact, lab exercises, ceremonies, and interpersonal interactions throughout the ROTC curriculum. MSL IV Cadets work alongside cadre members to serve as peer role models for junior Cadets.

- *Military History.* Cadets review vignettes and case studies, which provide opportunities for critical reasoning in evaluating tactics, leadership styles, problem solving, and decision making. MSL IV Cadets conduct military ethics case studies, a battle analysis, and participate in a staff ride to apply lessons learned from military history.

- *Management and Administration.* Cadets learn Army programs, policies, and procedures related to areas such as organization, human resources, management, administration, training, and facilities in order to support Army operations.

Tactics and Techniques

- Training and mentoring underclass Cadets to learn and lead *tactical operations* is the major focus of the two MSL IV courses. Cadets are expected to shift from mastering an understanding of tactical operations to teaching others to learn and master these operations through in-depth study and experiential leadership opportunities. MSL IV Cadets plan, prepare, and lead the labs in which Cadets are expected to develop and demonstrate a proficient understanding and ability to perform basic land navigation, troop leading procedures, and squad tactical operations.

MSL 402 Course Overview: Leadership in a Complex World

MSL 402 develops proficiency in planning, executing, and assessing complex operations, functioning as a member of a staff, and providing leadership-performance feedback to subordinates. Cadets are given situational opportunities to assess risk, make sound ethical decisions, and provide coaching and mentoring to fellow ROTC Cadets. MSL IV Cadets are measured by their ability to give and receive systematic and specific feedback on leadership abilities using the Socratic model of reflective learning. Cadets at the MSL IV level analyze and evaluate the leadership values, attributes, skills, and actions of MSL III Cadets while simultaneously considering their own leadership skills. Attention is given to preparation for success at BOLC II and III, and the development of leadership abilities. Cadets must meet the following objectives:

Apply leadership skills of coaching and counseling subordinate Cadets to prepare for future leadership roles.

- Facilitate learning through after action reviews and counseling sessions
- Conduct briefings and training meetings in accordance with Army standards
- Use Army Values and ethics to make decisions
- Apply counseling techniques to specific situations.

Demonstrate proficiency in following and enforcing Army policies and procedures.

- Demonstrate familiarity with Army customs and courtesies
- Demonstrate an understanding of the workings of a platoon command team
- Demonstrate a working knowledge of the Army's programs on Equal Opportunity, Prevention of Sexual Harassment (POSH), and Sexual Assault Prevention and Response (SAPR)
- Apply correct procedures in supply and maintenance.

Prepare personal developmental plan using the Junior Officer Developmental Support Form and Officer Evaluation Report model.

- Understand and prepare for the Combat Lifesaver Course
- Display fluency in conducting battle analysis, awareness of cultural differences and their effects on operations, and a firm grasp of terrorist culture
- Apply the principles of force protection and operational security
- Demonstrate good management of personal finances.

The Role of the MSL IV Cadet

LEADS. As potential Army officers, you will be challenged to study, practice, and evaluate adaptive team leadership skills as you are presented with the demands of preparing for your commission and attendance at BOLC II. Increasingly complex scenarios related to small-unit tactical operations are used to develop self-awareness and critical-thinking skills. You will receive systematic and specific feedback on your performance as a battalion staff officer by your cadre, just as commissioned officers serving on staffs receive counsel and feedback from their raters. You will be given numerous leadership opportunities as an MSL IV Cadet that will help you develop and improve your leadership skills. Taking these opportunities seriously and learning from your mistakes in a school environment will help build your confidence as a leader and provide you with better leader presence when you arrive at BOLC II and your first unit of assignment.

DEVELOPS. Learning the skills required of a competent officer and leader demands that you participate actively in learning through critical reflection, inquiry, dialogue, and group interactions. MSL 402 will teach you competency-based leadership described in FM 6-22 as it relates to your responsibility to earn your commission and to develop junior Cadets to become future Army lieutenants. Based on your understanding and experience of adaptive team leadership, you will work with subordinate Cadets to identify activities (such as club leadership, sports teams, event planning, or other extracurricular activities) in which they are able to practice adaptive leadership skills. You are also encouraged to continue in your own leadership roles beyond ROTC. While everyone is responsible for contributing to the success of the learning experience, your role as an MSL IV Cadet is to lead others in learning by offering constructive feedback and encouragement as a role model of leading and learning.

ACHIEVES. Extensive leadership preparation discussions and exercises are embedded throughout the MSL 402 course. MSL IV Cadets are encouraged to work together as a team and with their instructors in assessing training events, planning, coordinating and executing leadership labs and FTXs, constructing assignments, recommending courses of action (COAs) and agendas, raising questions for discussion, and providing feedback to MSL III Cadets who are learning to lead. Collaborative learning is enhanced when MSL IV Cadets challenge subordinate Cadets to describe adaptive team-leadership lessons learned from lab and FTX experiences. The bottom line is that as an MS IV, your PMS and cadre will expect you to perform and will expect positive results from you.

Academic Approach

The MSL curriculum is outcomes based and designed to focus on Cadet learning, rather than on any specific subject matter. Focusing on the Cadet requires student-centered objectives and conscious attention to how Cadets react to the instruction received. For effective instruction, Cadets need the opportunity to apply the knowledge received from instruction received by experienced cadre. Too often, academic instruction is limited to the delivery of information, either through reading assignments, lectures, or slide presentations. Active, student-centered learning, in contrast, is founded on the belief that interaction is central to the learning process. Learning occurs during class in the same way it does outside the classroom: through unstructured and structured experiences in which the Cadet interacts with cadre, with the instructional material, and with other Cadets. Helpful synonyms for ROTC's student-centered approach to learning are experiential learning, direct experience, discovery learning, experience-based learning, and participatory learning. All of these approaches center around five basic steps:

Helpful synonyms for ROTC's student-centered approach to learning are experiential learning, direct experience, discovery learning, experience-based learning, and participatory learning.

1. Readiness for and openness to the experience
2. The experience itself
3. Reflection upon the experience
4. Analysis, application of theory, or additional explanation of information to clarify the relationship between theory and actions, with an understanding of lessons learned regarding needed changes
5. The opportunity to re-experience (practice in new situations/practical exercises).

The emphasis must first be on the Cadet's pre-class preparation. Cadets must come to class with a foundation of knowledge from their pre-class readings. This allows the cadre to apply the Socratic model of reflective learning during the 50 minutes of classroom instruction. During this limited contact hour, the cadre can focus on explaining the concepts or material that needs clarifying.

How to Use This Textbook

The readings in this textbook have been compiled to prepare the Cadet to participate actively and productively in MSL classes and labs. The chapters are divided into four MSL curriculum tracks as follows:

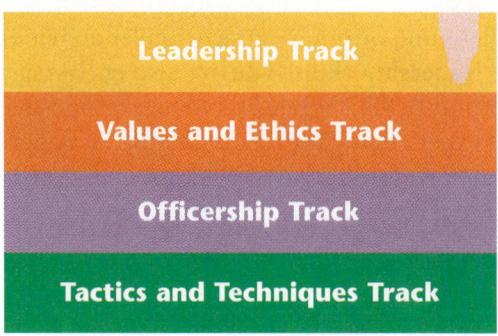

To be most effective, MSL class sessions are best sequenced to coincide with Leadership Lab schedules, which may vary from campus to campus due to weather, academic calendars, and other local variables. Thus, class sessions may not necessarily follow the same sequence as textbook chapters. *Cadets must follow the reading assignments given by their instructors to ensure they are adequately prepared for each class session.*

The first page of each chapter orients the Cadet to the key points to be covered in the reading assignment. At the end of each chapter, learning assessment questions serve as "checks on learning" for the Cadet to ensure he or she understands the key points of the chapter. Additionally, vignettes, scenarios, case studies, and critical-thinking questions are dispersed throughout the chapters to help the Cadet build critical-thinking skills and to apply the coursework to real-world situations. The learning assessment questions at the end of each chapter are aligned with the learning objectives for each coinciding lesson.

Cadet Resources

Cadet Textbook. This textbook contains the readings that support the MSL 402 course, Adaptive Leadership.

Blackboard (Bb). The Blackboard course web site, **http://rotc.blackboard.com**, contains MSL course materials.

CONCLUSION

The Basic Officer Leader Course (BOLC) common core task list forms the foundation of competencies a second lieutenant needs to know upon arrival at his or her first unit. Today's Army officer develops through a progression of BOLC sequential learning programs designed for pre-commissioning (BOLC I), common tactical training that is focused on warrior tasks and battle drills (BOLC II), and basic branch-specific training (BOLC III). The ROTC program is the implementation of BOLC I in a university setting. Today's ROTC Cadet represents the future leadership of our great nation. Such responsibility must be carried by officers well versed in the principles and practices of effective leadership, military operations, and personal development. A future officer must be a leader of character, of presence, and of intellectual capacity—a professional who is able to think critically and ready to lead Soldiers in the Contemporary Operating Environment (COE). The MSL IV year of ROTC forges this officer through a curriculum of leader development, officership, values and ethics, and personal development in preparation for the crucible event of BOLC II and the ultimate test in his or her first unit of assignment.

Although this course prepares you for this challenge, it is your responsibility to live the leader attributes while adopting and demonstrating Army Values at all times—both on and off campus. The qualities of an Army Officer are not words professed for an exam or exercise. At the MSL IV level, these qualities are the expression of a professional prepared to "support and defend the Constitution of the United States against all enemies, foreign or domestic." Your commitment to excellence in the Warrior Ethos, tactical proficiency, adaptive team leadership, and developing adaptive leaders is essential to the success of the Army of the future.

References

Cadet Command Reg. 145-3, ROTC Precommissioning Training and Leadership Development.

Field Manual 6-22, *Army Leadership: Competent, Confident, and Agile.* 12 August 2006.

Section

1

PLATOON COMMAND TEAM

Key Points

1 Rank Authority and Duty

2 Platoon Leadership Duties

3 Preparing for Your First Duty Station

The American soldier . . . demands professional competence in his leaders. In battle, he wants to know that the job is going to be done right, with no unnecessary casualties. The noncommissioned officer wearing the chevron is supposed to be the best soldier in the platoon and he is supposed to know how to perform all the duties expected of him. The American soldier expects his sergeant to be able to teach him how to do his job. And he expects even more from his officers.

General of the Army Omar N. Bradley

Introduction

Before long, you will receive your commission as an Army second lieutenant. You'll attend Basic Officer Leader Course (BOLC) II and BOLC III training and then become a platoon leader at your first duty station.

That platoon already exists. Its squad leaders and platoon sergeant are all Army veterans with several years' experience. They have likely served in Iraq or Afghanistan or both. They will be waiting for you to display the skills and knowledge the Army expects of its junior leaders. When you become the platoon leader, *Be, Know, Do* will be far more than a slogan you learned in ROTC—it will be the stuff of your daily life.

To help you prepare for this new challenge, this section will review the authority and rank structure in the Army, discussing the different roles of Department of the Army (DA) civilians, officers, noncommissioned officers (NCOs), and warrant officers. It will cover the roles and responsibilities of the platoon leader and the platoon sergeant. Finally, it will give you tips on how to prepare for your first assignment.

The success of any platoon depends heavily on the relationship between you, the young and inexperienced platoon leader, and your experienced and probably much older platoon sergeant. The two of you are the platoon command team. You must take this unique relationship very seriously and work hard at it.

The NCO as Mentor

mentor

a wise and trusted counselor or teacher

Senior NCOs have a great deal of experience that is valuable to officers. An officer who has an NCO as a **mentor** is taking advantage of that experience and of the unique perspective NCOs develop in leadership, training, and professionalism. Even very senior officers seek trusted NCOs' advice and counsel. A mentorship relationship that is unique in the Army and the NCO Corps is the relationship between a platoon sergeant and his [or her] young platoon leader. Especially in their early years, young officers need to be paired with senior experienced NCOs. The relationship that frequently comes from this experience tends to be instrumental in the young officers' development. Young officers may forget a lot of things about their time in the military, but they will never forget, good or bad, their first platoon sergeant.

Field Manual 7-22.7, *The Army Noncommissioned Officer Guide*

Rank Authority and Duty

You already know much about this topic, but it is worth reviewing the key points in preparation for your first assignment. You will work or interact in varying degrees with Department of the Army (DA) civilians, commissioned officers, warrant officers, and noncommissioned officers (NCOs). You should pay careful attention to the authority and duties of each Army member.

DA Civilians

As members of the executive branch of the federal government, DA civilians are an important part of the Army. They derive their authority from a variety of sources, such as commanders, supervisors, Army regulations, and Title 5 of the United States Code (USC). DA civilians' authority is job-related: They normally exercise authority connected with their positions. DA civilians fill positions in staff and base-sustaining operations that officers and NCOs would otherwise have to fill. Senior DA civilians establish policy and manage Army resources but do not have the authority to command.

The complementary relationship and mutual respect between the military and civilian members of the Army is a long-standing tradition. Since the Army's beginning in 1775, military and DA civilian duties have stayed separate, yet necessarily related. A combination of traditions, functions, and laws defines the particular duties of military and civilian members of the Army.

DA civilian positions have existed since the Army's beginnings in 1775.

Officers

Commissions are legal instruments the president of the United States uses to appoint and exercise direct control over qualified people to act as his or her legal agents and help him or her carry out his or her duties. The Army retains this direct-agent relationship with the president through its commissioned officers. The commission is the basis for an officer's legal authority. Commissioned officers command, establish policy, and manage Army resources. They are normally generalists who assume progressively broader responsibilities over the course of a career.

An officer:

- Concentrates on collective training, which will enable the unit to accomplish its mission
- Is involved primarily with unit operations, training, and related activities
- Concentrates on unit effectiveness and unit readiness
- Pays particular attention to the standards of performance, training, and professional development of officers as well as NCOs
- Creates conditions—makes the time and other resources available—so the NCO can do the job
- Supports the NCO.

Warrant Officers

Warrant officers are highly specialized, single-track specialty officers who receive their authority from the secretary of the Army upon their initial appointment. Title 10 USC, however, authorizes the commissioning of warrant officers (WO1) upon promotion to chief warrant officer (CW2). Like commissioned officers, these commissioned warrant officers are direct representatives of the president of the United States. They derive their authority from the same source as commissioned officers but remain specialists—in contrast to commissioned officers, who are generalists.

Warrant officers can and do command detachments, units, activities, and vessels as well as lead, coach, train, and counsel Soldiers. As leaders and technical/tactical experts, warrant officers provide valuable skills, guidance, and expertise to commanders and organizations in their particular specialty.

Warrant officers also provide mentorship, leadership, and training to NCOs to support technical, tactical, and mission-related tasks. The relationship of the warrant officer to the NCO is similar to that between the NCO and the commissioned officer. They rely on each other for help, advice, and assistance to accomplish the unit's mission.

A warrant officer:

- Provides quality advice, counsel, and solutions to support the command
- Executes policy and manages Army systems
- Commands special-purpose units and tasks—organizes operational elements
- Focuses on collective, leader, and individual training
- Operates, maintains, administers, and manages the Army's equipment, support activities, and technical systems
- Concentrates on unit effectiveness and readiness
- Supports the NCO.

Noncommissioned Officers

NCOs are "the backbone of the Army"—they train, lead, and take care of enlisted Soldiers. They receive their authority from their oaths of office, law, rank structure, Army traditions, and regulations.

This authority allows them to direct Soldiers, take actions required to accomplish the mission, and enforce good order and discipline. NCOs represent officers and, sometimes, DA civilian leaders. They ensure that their subordinates are prepared to function as effective unit and team members. While commissioned officers command, establish policy, and manage resources, NCOs conduct the Army's daily business.

The chain of command backs up the **NCO support channel** by legally punishing those who challenge the NCO's authority. But it does so only if the noncommissioned officer's actions and orders are sound, intelligent, and based on proper authority.

In training, the NCO's focus is on individual Soldiers, whereas you, as platoon leader, concentrate on collective or unit training.

A noncommissioned officer:

NCO support channel

a leadership chain or channel of communication and supervision from the command sergeant major to first sergeant and then to other NCOs and enlisted personnel of the units

- Conducts the daily business of the Army within established orders, directives, and policies
- Focuses on individual training and leading Soldiers and teams
- Ensures each subordinate is well trained, highly motivated, ready, and functioning
- Follows orders of officers and NCOs in the support channel
- Gets the job done.

Platoon Leadership Duties

The platoon's leadership consists of you as platoon leader and your senior NCO as platoon sergeant.

Platoon Leader

As platoon leader, you are responsible for everything your platoon accomplishes or fails to accomplish. You lead your unit, establish policy, and plan and program the unit's work. Your focus is on collective training, unit operations, and the unit's overall effectiveness. In short, you must create conditions that enable the platoon to accomplish its mission. You must:

- Set the example and standards
- Lead the platoon in support of company and/or battalion missions
- Inform your higher commander of your actions when operating without orders
- Plan with the help of key team members
- Request support for the platoon when needed
- Assist your platoon sergeant in planning and coordinating the platoon's combat service support (CSS) efforts
- Receive on-hand status reports from platoon sergeant, squad leaders, and others
- Review platoon requirements based on tactical plans
- Check the work of the platoon sergeant and squad leaders
- Ensure your Soldiers' workload is reasonable
- Build a cohesive team
- Develop and support your platoon sergeant and the NCO support channel.

Platoon leaders are responsible for all the platoon's accomplishments and failures.

Platoon Sergeant

The platoon sergeant is the senior NCO in the platoon and second in command. He or she typically has 12 to 18 years of experience and serves as your primary assistant and adviser, leading the platoon in your absence. The platoon sergeant:

- Executes the daily business of the platoon within the parameters of orders, directives, and policies
- Supervises the platoon's administration, logistics, and maintenance
- Oversees Soldiers' training on both individual and collective (team/crew) tasks
- Ensures Soldiers' needs are met
- Assists in training the platoon leader (you)
- Serves as platoon leader in your absence.

Expectations of the Platoon Command Team

You should expect the following from your platoon sergeant:

- Loyalty, honesty, and trust
- Commitment to and enforcement of Army Values
- Technical and tactical proficiency
- A take-charge attitude
- Sound and timely decision making
- Care of Soldiers and concern for their well-being and professional development
- Sharing of his or her knowledge and experience with you without being condescending or disrespectful.

The Army Values—
LDRSHIP
Loyalty
Duty
Respect
Selfless Service
Honor
Integrity
Personal Courage

In return, your platoon sergeant will expect from you:

- Loyalty, honesty, support, and trust
- Commitment to and enforcement of Army Values
- Dedication to the unit, its mission, and the welfare of its Soldiers
- Commitment to making training a high priority in the unit
- Flexibility and adaptability
- Forward thinking and planning—setting the conditions for success
- Sound and timely decision making
- Excellent oral and written communication skills
- Integrity and candor (forget the politics)
- Willingness to listen and eagerness to learn.

The Platoon Leader–Platoon Sergeant Relationship

The relationship between you and your platoon sergeant is one of the most important in the Army. It is essential to the platoon's effective operation. You must work to build a relationship characterized by:

- *Mutual respect and trust.* A positive relationship cannot exist without these qualities. Once trust is lost, it is almost impossible to recover.
- *Purposeful redundancy of responsibilities.* While it can be a potential source of friction if you don't manage it properly, an overlap of responsibilities is necessary. Because the platoon sergeant must always be ready to step in and lead the platoon, he or she must be comfortable operating in the boss's—in your—world.
- *Effective communication.* The platoon sergeant must understand the mission and your intent, as well as that of the company and battalion, if he or she is to make and implement decisions and accomplish the mission. Similarly, to make sound decisions you must know the morale, readiness, and abilities of your Soldiers.

Here's the bottom line—your platoon's effectiveness in garrison and in combat is directly related to the quality of your relationship with your platoon sergeant.

Critical Thinking

How do you earn your platoon sergeant's respect? Do your NCOs have to *like* you to respect and trust you?

Show an Interest in Your Soldiers

I think the interest of [a Soldier's] immediate supervisor, be it a sergeant, a lieutenant, whoever it happens to be . . . the more interest they show in him and his welfare the better he responds. . . . You show me the man who is reluctant to fight, tries to stay away from the combat situation and I'll show you somebody that has a leader who doesn't take very much interest in his men. He doesn't work at finding out the strengths and weaknesses of his own platoon or of his own company.

LTC Douglas S. Smith

Noncommissioned officers are the backbone of the Army and the reason our Army is the best trained, most professional, and most respected in the world.

GEN Erik K. Shinseki and Jack L. Tilley, Sergeant Major of the Army
Field Manual 7-22.7, *The Army Noncommissioned Officer Guide*

Critical Thinking

Why are NCOs "the backbone of the Army"? What does that say about your relationship with your platoon sergeant and squad leaders?

Preparing for Your First Duty Station

When you receive orders to report to your first duty station, what should you do to prepare?

First, you must understand your duties, your platoon sergeant's duties, and the relationship between the two of you. The paragraphs earlier will help you do that. In addition, BOLC II and III will help you polish your skills and give you new ones. You should also increase your proficiency in key areas where you may be weak and obtain relevant information about your new unit and post.

Increase Your Proficiency

There are a number of things you can do to increase your knowledge and skills:

- Attend additional military schools such as Ranger School, Airborne School, and others
- Strengthen known weaknesses. If you know you could do better in land navigation or physical fitness, seek additional training
- Read! Before you show up at your platoon, know the equipment you will have
- Read the Army manuals on it, pay attention in BOLC weapons/equipment maintenance training. Read all the field manuals you can find that relate to anything your platoon does
- Seek advice. Talk to platoon leaders you may already know, such as officers in the class before yours. Talk to the officers and NCOs you have met in ROTC activities on campus
- Go to *http://platoonleader.army.mil* and see what other platoon leaders are saying.

Actions to Take Before You Report

Other than improving yourself and searching for general information, you can take other actions before you show up for duty. It's better to do what you can early—you won't have a lot of time once you hit the ground in your first assignment.

- Prepare and send a letter of introduction to your future commander
- Obtain as much information about your new location and unit as you can
- Call the company executive officer to get advice
- Make checklists, a leader's book, and other support documents to use as aids in carrying out your duties
- Draft a plan to take charge.

What to Do When You Get There

Once you arrive at your unit, the Soldiers will be looking to you as their leader. Additionally, both your commander and your peers will expect certain things from you.

According to the *Army Officer's Guide,* here are some things you should do when you arrive at your station:

- Make a good first impression
- Show your commander and Soldiers respect
- Let your platoon sergeant know you want his or her advice
- Let your platoon sergeant know you support him or her
- Don't make changes until you have observed for a while
- Don't criticize the previous leader.

Information You Should Find Out

The *Army Officer's Guide* further recommends that you ask the following questions when you arrive at post:

- What is the unit's mission?
- What are the current training or work projects?
- What reports are required?
- Who does what?
- How is classified material handled here?

Here are some other steps any leader should take when arriving at a new assignment:

- Meet with your new commander and obtain his or her command philosophy, goals, objectives, and plans
- Determine what is expected of you, including (if appropriate) completion of your Officer Evaluation Report Support Form and Developmental Support Form (DA Form 67-9-1 and DA Form 67-9-1A)
- Determine what higher headquarters expects of your platoon
- Introduce yourself to the members of your new platoon
- Inventory and sign for platoon property and have those who use specific platoon equipment sign for it from you
- Meet and talk with the outgoing platoon leader
- Read key documents, such as mission statements, standing operating procedures (SOPs), policy memorandums, and authorization documents
- Determine how your mission fits into the mission of the next higher headquarters
- Identify the key people outside your platoon whose support you need to accomplish your mission
- Determine the functions you are responsible for, such as training, maintenance, and administration
- Determine the resources available to help you accomplish the mission
- Determine who reports directly to you
- Determine the current state of the platoon's morale.

Critical Thinking

What should you do to make sure you create a good first impression when arriving at your first duty station?

CONCLUSION

Be prepared for your first assignment. Increase your skills where necessary, take appropriate actions, and get the information you need to do well from day 1.

Above all, pay attention to relationships. Remember from your leadership training that your leadership style depends as much on your Soldiers as it does on you. Starting with your first assignment, you will need to understand the roles of the people you will work with—DA civilians, warrant officers, commissioned officers like yourself, and NCOs. Perhaps your most important relationship will be with your platoon sergeant. Understand clearly your own duties as well as his or hers, including where they overlap, and have a plan for developing the kind of bond that will stand up even under great pressure. Getting started right will help you immensely in accomplishing your goals—and your platoon in accomplishing its mission.

Key Words

mentor
NCO support channel

Learning Assessment

1. Describe the duties, responsibilities, and authority of officers, warrant officers, noncommissioned officers, and DA civilians.

2. Explain the duties and responsibilities of a platoon leader.

3. List the duties and responsibilities of a platoon sergeant.

4. Describe actions you should take upon arrival at your new assignment.

5. Explain actions you should take in preparation for taking charge of a platoon.

References

AR 600-20, *Army Command Policy*. 18 March 2008.

Bonn, K. E. (2002). *Army Officer's Guide*. 49th Edition. Mechanicsburg, PA: Stackpole Books.

Exit Interview With LTC Douglas S. Smith, Commander, 2d Battalion, 47th Infantry, 9th Infantry Division, VNIT 457. (1 July 1969). Department of the Army, Center for Military History. Retrieved 27 September 2005 from http://www.army.mil/cmh-pg/documents/vietnam/vnit/vnit457.htm

Field Manual 3-21.8, *The Infantry Rifle Platoon and Squad*. 8 March 2007.

Field Manual 6-22, *Army Leadership: Competent, Confident, and Agile*. 12 October 2006.

Field Manual 7-22.7, *The Army Noncommissioned Officer Guide*. 23 December 2002.

Training Task 158-100-1333. Take Charge of a Company, Staff Section or Similar Sized Organization. (1999). US Army. Retrieved 27 September 2005 from http://www.atsc.army.mil/itsd/comcor/cg1333s.htm

EQUAL OPPORTUNITY AND PREVENTING SEXUAL HARASSMENT AND ASSAULT

Key Points

1 Army Equal Opportunity (EO)

2 Prevention of Sexual Harassment (POSH)

3 Sexual Assault Prevention and Response (SAPR)

4 Case Studies

A single incident of sexual assault can impact a unit's cohesion and readiness.

Consideration of Others Handbook

Introduction

As you have learned in your ROTC studies, the Army is a values-based institution. Army Values reflect the culture of American society and the ideals of the profession of arms.

One of your responsibilities as a platoon leader will be to instill in each of your Soldiers an awareness that one Soldier's attitudes, actions, and words affect all the others in your unit. Soldiers' willingness to take responsibility for those attitudes, actions, and words—to the point of changing them when necessary—is what the Army's Consideration of Others (CO2) philosophy is all about. Three major elements of CO2 are equal opportunity (EO), prevention of sexual harassment (POSH), and responding to and preventing sexual assault.

All officers and Soldiers must understand that the Army has zero tolerance for sexual harassment and sexual assault. The best approach to dealing with both is to head them off. Learning to identify situations that foster or ignore such behaviors is critical to preventing them. The following vignette from the *Consideration of Others Handbook* demonstrates the type of situation that can have serious consequences for the parties involved as well as for the unit.

SGT Green and SGT Taylor

SGT Green and SGT Taylor take weight training classes together, and both are well respected in their unit. SGT Green asked SGT Taylor to the movies; this was their first date. Following the movie, SGT Green and SGT Taylor went back to SGT Taylor's apartment to watch the late show on television. While watching television, SGT Green put his arm around SGT Taylor's shoulder. A few minutes later, he put a hand on her breast. SGT Taylor moved his hand away, smiled, and said, "I don't know if we should do this." SGT Green began to kiss SGT Taylor and then lifted her skirt. SGT Taylor tried to squirm away from SGT Green and began protesting loudly, but he ignored her as he pushed her onto the couch and completed the act of intercourse. After SGT Green left the apartment, SGT Taylor contacted the Military Police to report a sexual assault.

Consideration of Others Handbook

Critical Thinking

Did SGT Green sexually assault SGT Taylor? Why might SGT Green have ignored SGT Taylor's protests? What else might SGT Taylor have done or said to protect and assert herself?

Army Equal Opportunity (EO)

Equal opportunity (EO) rests on the premise that your Soldiers have a right to excel regardless of race, color, creed, gender, ethnic group, religion, or national origin. The Army supports an entire system of training and policy support dedicated to promotion of equal opportunity. Your job is to use these resources to educate your Soldiers about racial, ethnic, and religious groups that differ from their own and about the effects of their own actions, attitudes, and words upon Soldiers from these differing groups.

Two Types of EO Complaints

The Army's EO complaint processing system addresses complaints that allege unlawful discrimination or unfair treatment based on race, national origin, color, gender, or religious affiliation. It also addresses issues of sexual harassment. In dealing with EO complaints, it's important that you always attempt first to solve the problem at the lowest possible level within the organization.

The Army recognizes two types of EO complaints within its EO complaint process: *informal* and *formal*.

An *informal* complaint is any complaint that a Soldier, family member, or Department of the Army (DA) civilian does not wish to file in writing. The individual may be able to resolve informal complaints directly, with the help of another unit member, the commander, or another person in the complainant's chain of command. Typically, these informal issues can be resolved through discussion, identifying the problem, and clarifying the issues. An informal complaint is not reportable.

A *formal* complaint is one that a complainant files in writing, in which the Soldier swears to the accuracy of the information. Formal complaints require specific actions, are subject to timelines, and require documentation of the actions taken.

Soldiers have 60 calendar days from the date of the alleged incident in which to file a formal complaint. This time limit sets reasonable parameters for the inquiry or investigation and resolution of complaints, and it ensures the availability of witnesses, accurate recollection of events, and timely remedial action.

Alternative Agencies

Although the Army strongly encourages handling EO complaints through the chain of command, this is not the only channel. Should the Soldier feel uncomfortable in filing a complaint with the immediate chain of command, or should the complaint be against a member of the chain of command, the Army provides a number of alternative agencies to handle the complaint. Soldiers may also contact:

- The equal opportunity adviser
- The chaplain
- The provost marshal
- The staff judge advocate (SJA)
- The housing referral office (for complaints of discrimination in off-post housing)
- The inspector general.

In addition to the alternative agencies, each base or post has an EO hotline. While this hotline provides information on discrimination and sexual harassment, EO complaints cannot be filed over the phone. You must post the EO hotline phone number and ensure that everyone in your command knows where to find it.

The Right to Appeal

A Soldier may appeal to the next higher commander in the chain of command if he or she believes an investigation was faulty or that the actions his or her supervisor took did not resolve the complaint. The Soldier filing the complaint may not appeal any action taken against the perpetrator, however. The Soldier must file the appeal within seven calendar days after notification of the results of the investigation. Once the Soldier initiates the appeal, the commander has three calendar days to refer the appeal to the next higher unit commander. The commander to whom the appeal is made has 14 calendar days to review the case and act on the appeal (i.e., approve it, deny it, or conduct an additional investigation).

Prevention of Sexual Harassment (POSH)

The Army has had to deal with several high-profile sexual-misconduct cases in recent years. In 1996, several female recruits at the Aberdeen Proving Ground complained of sexual harassment by drill instructors. A drill sergeant testified that he was part of a sex ring targeting female trainees. The Army eventually charged 11 sergeants and one captain at Aberdeen with offenses ranging from rape to adultery to obstruction of justice. Several received prison terms. The base commander, MG Robert Shadley, was given a letter of reprimand and later retired from the Army. LTC Martin Utzig, who commanded seven of the accused drill sergeants, was suspended.

Sexual Harassment Behaviors

• *Verbal*
  – *Jokes, sexually explicit profanity, describing physical appearance, terms of endearment*

• *Nonverbal*
  – *Staring, licking lips suggestively*
  – *Displaying sexually explicit pictures or screen savers*
  – *Sexually oriented e-mail, notes, printed material, etc.*

• *Physical*
  – *Touching, patting, pinching, blocking passage*
  – *Sexual assault may be an extreme form*

In 1998, an Army investigator ruled that Sergeant Major of the Army Gene McKinney should be court-martialed on 22 counts of sexual misconduct. The charges stemmed from allegations by six women who had worked for him. McKinney was convicted of obstruction of justice.

Retired MG David Hale was convicted in 1997 of having affairs with the wives of subordinate officers. Hale had been appointed deputy inspector general and served for four months before he was allowed to retire after a complaint was filed against him. The Army called him out of retirement to charge him; after he pleaded guilty to eight charges, the Army demoted him to brigadier general. A year later, MG John Maher III, former commander of the 25th Infantry Division, was demoted to colonel and forced to retire for having an affair with subordinates' wives.

In another case, LTG Claudia Kennedy, the Army's first female three-star general, accused MG Larry Smith of groping her in her Pentagon office in 1996. She filed a formal complaint after Smith was appointed deputy inspector general in 2000. The position would have made him responsible for investigating wrongdoing in the service, including sexual-harassment allegations.

Female Soldiers compose a significant percentage of the active Army today. Clearly, the Army cannot be combat ready if its male and female Soldiers cannot work together.

In recent years, the issues of sexual harassment and sexual assault have received significant media and political attention in both government and private sectors. This heightened awareness of the causes of sexual harassment and sexual assault has intensified the national debate on prevention. The elimination of sexual harassment and sexual assault remains an Army goal.

Sexual harassment and sexual assault affect everyone. A positive working climate that promotes individual growth and teamwork is vital to your unit's combat readiness. Sexual harassment and sexual assault will destroy trust, teamwork, and ultimately, combat readiness.

As an Army leader, you must ensure that neither your perspective nor those of your unit members hinder the ability of men and women to work together in a professional manner. Your approach should combine education about the roles and status of male and female Soldiers in today's Army with lessons to make Soldiers aware of and more sensitive to the different ways men and women perceive things and communicate.

Sexual harassment and sexual assault victimize males as well as females and can occur at any time, on or off the job. The Army cannot and will not tolerate sexual harassment or sexual assault in any place or at any time.

What Is Sexual Harassment?

Sexual harassment is a form of gender discrimination that involves unwelcome sexual advances, requests for sexual favors, and other verbal or physical conduct of a sexual nature. A victim of sexual harassment usually feels that he or she must submit to such conduct as a term or condition of employment or risk his or her career in order to reject the harassment. This conduct has the purpose or effect of unreasonably interfering with an individual's work performance and could create an intimidating, hostile, or offensive work environment.

Any person in a supervisory or command position who uses or condones implicit or explicit sexual behavior to control, influence, or affect the career, pay, or job of a Soldier or civilian employee is engaging in sexual harassment.

Similarly, any Soldier or civilian employee who makes deliberate or repeated unwelcome verbal comments, gestures, or physical contact of a sexual nature is also engaging in sexual harassment.

Types of Sexual Harassment

Soldiers must have a clear understanding of the types of behavior that constitute sexual harassment. Two of these are *quid pro quo* and *hostile environment*. *Quid pro quo* refers to conditions a harasser places on a person's career or terms of employment in return for sexual favors. A *hostile environment* develops when Soldiers or civilians are subjected to offensive, unwanted, and unsolicited comments and behavior of a sexual nature.

Impact Versus Intent

Soldiers and civilians must understand that what they may consider to be joking or horseplay must be evaluated on its appropriateness and offensiveness as *perceived by the recipient*. When a person receives attention of a sexual nature that she (or he) neither wants, initiates, nor solicits, it is *unwelcome*. In determining whether such behavior constitutes sexual harassment, a primary concern is the *impact of the act upon the victim*, not the alleged harasser's intent. An excuse such as "I was only joking" is irrelevant.

"Reasonable Person" and "Reasonable Woman" Standards

The reasonable person standard or the reasonable woman standard is a measurement of the impact or expected reaction to sexual harassment. These standards are used to predict how a recipient reacts to or is affected by offensive behaviors—in other words, how would a "reasonable woman" or a "reasonable person" react to or be affected by this behavior? They ensure adequate sensitivity to a person's feelings and perspective while avoiding extremes. The purpose of adopting a reasonable woman standard is to avoid the issue of male bias that could exist in a reasonable person standard.

Reporting Sexual Harassment

All Soldiers and their family members have the right to prompt and thorough redress of complaints about sexual harassment without fear of intimidation or reprisal. AR 600-20 contains detailed information on the Army's EO complaint process.

The immediate chain of command is the primary channel for handling and correcting allegations of sexual harassment. Although a number of alternative channels are available, encourage Soldiers to bring their complaints to you for resolution at the lowest possible level. Should Soldiers feel uncomfortable in bringing their concerns to the chain of command or if the allegation of sexual harassment is against a member of the chain, a number of alternate agencies are available to assist in the complaint process.

Victims may file formal or informal complaints of sexual harassment. An informal complaint is one in which the complainant does not wish to file his or her grievance in writing. In attempting to resolve the problem at the lowest possible level, it may not be necessary to involve the commander or other members of the chain of command.

Complainants have 60 calendar days from the date of the alleged incident in which to file a formal complaint of sexual harassment. The Soldier filing the complaint is responsible for providing all pertinent information, including a detailed description of the incident and the names of witnesses and other involved parties. The commander who receives the complaint has 14 calendar days to resolve it or provide written feedback to the complainant. In special circumstances, the complainant may receive an extension of 30 additional calendar days.

Should the complainant disagree with the findings or actions taken to resolve the complaint, he or she may file an appeal within seven calendar days after notification of the final resolution of the complaint. The complainant files the appeal with the commander who processed the complaint, the next higher commander within the chain, or with the commander who has the authority to convene a general court-martial.

Should a complainant feel that he or she is a victim of intimidation or reprisal actions, he or she must immediately report such incidents to the chain of command or other alternate agencies.

Victim Impact

Soldiers must understand the devastating effects sexual harassment can have on a victim and on unit readiness. Problems from sexual harassment can manifest themselves in a number of ways. Some are obvious, while others may be invisible. The first and most obvious impact sexual harassment has on victims is that it interferes with their work performance. A Soldier who has to fend off offensive and repeated sexual advances cannot perform quality work.

Sexual harassment also creates a hostile environment by placing unreasonable stress on the victim. The impact of this form of stress on the victim can be devastating. It can affect not only the victim's ability to perform effectively on the job, but can also have an adverse effect on off-duty time.

Sexual harassment also puts a high degree of fear and anxiety into the workplace. When the harassment is *quid pro quo*, the victim's fear of losing job or career opportunities can undermine a unit's teamwork and morale. Sexual harassment propagates a negative form of stress that can affect everyone in the workplace.

The bottom line is this: Anyone who is sexually harassed will be less productive, and the unit climate will likely suffer. Soldiers and civilian employees can reach their full potential only in an environment that fosters personal dignity and mutual respect.

Sexual Harassment Checklist

- Is the behavior inappropriate for the workplace?
- Is the behavior sexual in nature or connotation?
- Is the conduct unwanted, unwelcome, or unsolicited?
- Do the elements of power, control, or influence exist?
- Does the situation indicate a *quid pro quo* relationship?
- Does the behavior create a hostile or offensive environment?
- Is the behavior repeated as it relates to gender treatment?
- How would the behavior affect a "reasonable person" or "reasonable woman?"

Sexual Assault Prevention and Response (SAPR)

Sexual assault—forcing someone to engage in sexual activity without her or his consent—*is a crime.* Sexual assault includes rape, nonconsensual sodomy (oral or anal sex), indecent assault (unwanted, inappropriate sexual contact or fondling), or attempts to commit these acts. Sexual assault can occur without regard to the gender of the parties involved, the marital status, or the age of the victim.

The issue of *consent* is crucial to sexual assault. *Consent* doesn't mean only the victim's failure to offer physical resistance. There can be no consent when a perpetrator uses force, threat of force, coercion, or when the victim is asleep, incapacitated, or unconscious.

Sexual acts that do not meet the definition of sexual assault may still violate the Uniform Code of Military Justice (UCMJ). Examples of other sex-related offenses could include indecent acts with another Soldier and adultery.

> **sexual assault**
>
> *intentional sexual contact characterized by use of force, physical threat, or abuse of authority, or when the victim does not or cannot consent—sexual assault can occur without regard to gender, spousal relationship, or age*

Sexual Assault and Sexual Harassment

It's important to recognize that sexual assault and sexual harassment are not the same, although they are related. Sexual assault refers specifically to rape, forcible sodomy, indecent assault, and carnal knowledge as defined by the UCMJ. Sexual assault must involve physical contact and is a crime. While sexual harassment may involve physical contact, it can also refer to verbal or other forms of gender discrimination of a sexual nature.

Both sexual assault and sexual harassment are incompatible with Army Values and with the Warrior Ethos. Both sexual assault and sexual harassment detract from a positive unit climate and will have detrimental effects on individual growth and teamwork. Teamwork is vital to combat readiness.

The Army actively uses training, education, and awareness to prevent sexual assault; promotes the sensitive handling of sexual assault victims; offers them confidential counseling; and holds those who commit sexual assault offenses accountable to the law.

Army policy demands that all victims of sexual assault be treated with dignity, fairness, and respect. The Army takes every sexual assault incident seriously, thoroughly investigates the incident, and holds accountable those who commit offenses.

Preventing Sexual Assault

The Army holds you responsible as a leader and as a Soldier to work to prevent sexual assault in your unit. You accomplish this by creating a climate in your unit that promotes safety, educates your Soldiers on how to reduce the risk of sexual assault, and makes everyone feel free to report incidents.

A major part of the Army's effort to prevent sexual harassment and sexual assault is to provide training and education for Soldiers on these types of behaviors. As a platoon leader, you should take the following training actions to help reduce the risk of sexual assault in your unit:

- Educate and train your unit on sexual assault prevention
- Educate Soldiers about the definition of sexual assault, the Army policy regarding sexual assault, and prevention measures they can take to reduce their risk of sexual assault
- Conduct Consideration of Others (CO2) training in your unit to increase Soldiers' understanding of the risks of sexual assault and the steps they can take to reduce the risk
- Conduct unit refresher training on sexual assault prevention in your unit (Soldiers will have already received sexual assault prevention training as part of their Initial Entry Training)
- Consider the risk of sexual assault and conduct unit safety briefs during high-risk periods, such as the holidays and deployments
- Monitor the unit climate to ensure that you and your noncommissioned officers (NCOs) are supportive of victims
- Ensure that Soldiers feel comfortable reporting sexual assault to you. You can do this by communicating your intention to protect victims of sexual assault and by making it clear that you will follow Army policy in fully investigating all incidents of sexual assault
- Communicate to your Soldiers that you will provide caring assistance to victims of sexual assault
- Make sure your Soldiers know that you will take whatever appropriate disciplinary action you can
- Continually assess the climate regarding the risk of sexual assault in your unit and mitigate or eliminate any factors that may promote sexual assault
- Demonstrate through your words and actions that sexual assault is unacceptable and incompatible with your values, Army Values, and the Warrior Ethos—tell your Soldiers through your words and deeds that you and the Army will not tolerate such behavior.

Reducing the Risk of "Date Rape" in Your Unit

Most victims know their rapist. According to the Rape, Abuse & Incest National Network (RAINN), about two-thirds of sexual assault victims in the United States know their assailants. *Acquaintance rape*, which includes date rape, refers generally to rapes that occur between two or more people who know one another. *Date rape* is a more specific term that refers to situations in which person A has consented to go on a date with person B, and person B then rapes person A. Both situations involve sexual assault because the victim does not consent.

A particular danger is sexual predators' use of "date rape drugs." Date rape drugs leave the victim helpless to stop a sexual assault. The perpetrator usually puts the drug in the victim's drink. Victims may be physically helpless, unable to refuse sex, and unable to remember what happened.

Definition of Consent

Consent shall not be deemed or construed to mean the failure by the victim to offer physical resistance. Consent is not given when a perpetrator uses force, threat of force, coercion, or when the victim is asleep, incapacitated, or unconscious.

Date rape drugs often have no color, smell, or taste and are easily added to flavored drinks without the victim's knowledge. Alcohol can worsen the drug's effects.

To prevent acquaintance or date rape within your unit you can:

- Educate Soldiers on the importance of maintaining their alertness by avoiding alcohol and drug use
- Encourage assertiveness and communication in dating and other intimate situations
- Educate Soldiers and Army civilian employees on the dangers of date rape drugs.

Reducing Risk in Deployed Units

You must be especially prepared and alert to the dangers of sexual assault when your unit is deployed. Some Soldiers on deployment adopt the mindset that the normal rules of everyday life are suspended, which heightens the risks for Army personnel in deployed environments. For one thing, sleeping areas (tents, bunkers, and other buildings) may be less secure.

Here are some tips for ensuring that your Soldiers remain safe on deployment:

- Encourage Soldiers to report any unauthorized males or females in sleeping areas
- Enforce security measures around sleeping areas, especially at night
- Monitor and control non-Army personnel present in deployed unit and working areas
- Encourage Soldiers to be alert and aware of their surroundings at all times
- Recommend that your Soldiers minimize their risk by traveling with a buddy.

Remember that other cultures may treat females very differently from the United States. This may mean that normal and acceptable behavior for an American woman may be perceived by a local man as having sexual meaning. Encourage Soldiers, especially women, to be assertive and clearly state if they feel uncomfortable with how someone is treating them in a foreign environment. Encourage them to report any inappropriate behavior to you immediately.

When Sexual Assault Occurs

Army leaders play a key role in the Army's response to sexual assault. These leaders include junior officers, commanders, supervisors, law enforcement personnel, legal and social services, and health-care personnel.

As an Army platoon leader, you are charged with "frontline" enforcement of the Army policy on sexual assault and with making sure your subordinates enforce it, too.

As an Army officer, you *must* report immediately any activity that indicates a sexual assault may take place, may have taken place, or has taken place. Report the activity or incident to your commander, Military Police, or another authority.

Immediately after a victim reports an assault to authority, you must follow established Army guidelines. These include:

AR 27-10 contains important information about victims' rights and the assistance available to victims.

- Don't blame the victim on the basis of his or her past
- Don't assume that the victim instigated the incident
- Inform each party of the victim's rights under AR 27-10
- Report the allegations to law enforcement for a thorough investigation
- Keep all information confidential and disclose information only to those who have an official need to know—it's the right of the accuser and the accused
- Notify the chaplain if the victim wants pastoral counseling or assistance
- Ensure that the needs of the victim's family are considered
- Make sure the victim is aware of the military and civilian resources available under the Victim and Witness Assistance Program (VWAP)
- Encourage the victim to get a medical examination, even if the incident occurred more than 72 hours ago. It is important for the victim to seek medical attention to assess possible injury, sexually transmitted diseases, and/or pregnancy.

As a commander, you have a responsibility to ensure that victims of sexual assault receive sensitive care and support and are not victimized a second time as a result of reporting the incident.

You have a range of command options available to help you fulfill your responsibility to protect sexual assault victims. You can geographically separate the victim and the alleged offender. You should determine whether the victim wants a transfer to another unit. By considering the victim's preferences and all relevant facts and circumstances of the case, you can avoid subjecting the victim to the "double victimization" sometimes perceived when a victim is transferred from the unit.

Military Protective Orders (MPOs), DD Form 2873, referred to as "no contact orders," are also an effective tool for you to maintain the victim's safety. You should seek the assistance of your judge advocate.

Some Army sexual assault victims report that they are hesitant to report sexual assaults when they feel they will bring disciplinary action upon themselves for offenses such as drug or alcohol use that may be related to the assault. As a commander, you can delay action on any victim misconduct related to an assault until after the investigation and prosecution for the assault is complete.

Sexual Assault Against Men

Men can also be victims of sexual assault. According to Defense Department statistics, as of 2006 about 3.5 percent of military men had experienced a sexual assault during their career. In 2002, 1 of every 8 American rape victims was male. Women are clearly not the only victims of sexual assault.

Sexual assault is devastating to all victims, male or female. Men can face special difficulties as victims of sexual assault, however. If a heterosexual man reports being sexually assaulted by another man, others may assume that the victim is homosexual.

Since most sexual assault victims are women, it can be especially difficult for men to report a sexual assault. For example, men may fear that their masculinity may be questioned if they report a sexual assault. Men may also fear that authorities or others will not believe their report.

Reporting Sexual Assault

Dr. David Lisak, a professor of psychology at the University of Massachusetts-Boston and director of the Men's Sexual Trauma Research Project, has conducted extensive research on men who commit sexual assaults. His research has shown that most sexual assaults are committed by a small number of men who perpetrate multiple offenses against victims with whom they have some degree of acquaintance. The implication from such research is that reporting sexual offenders can prevent or reduce additional crimes.

Soldiers should report any of the following activities immediately to you, the Military Police, or another authority:

- Someone planning to commit a sexual assault
- Conversations with others about getting another person drunk or "stoned" to make them less inhibited or easier to force into sexual relations
- Someone describing or bragging about a situation in which he or she physically forced another person into sex
- Conversations where someone brags that his or her partner didn't want to have sex, but they did so anyway
- Evidence of, or conversation about, the use of date rape drugs.

Remember: The safety of your Soldiers, your unit, and your community may depend on reporting of these incidents. Soldiers should report any suspicious behavior immediately.

Victims should also report sexual assault immediately. Any witnesses to the assault should report to the chain of command or a law enforcement agency—as a friend, acquaintance, family member, or fellow Soldier.

A sexual assault victim can file two types of reports. She or he always has the option as to which type of report to file and whom to file it with:

- A *restricted report* does not trigger the official investigative process. It allows a Soldier who is a sexual assault victim to disclose the details of her or his assault to specifically identified individuals and receive medical attention and counseling. To initiate a restricted report, the victim should confidentially report to a sexual assault response coordinator (SARC), a victim advocate, a health-care provider, or a chaplain.

- An *unrestricted report* triggers an official investigation. If a victim wishes to file an unrestricted report, she or he may give permission to one of the four restricted reporting resources she or he has chosen to make an unrestricted report. She or he may also report the assault directly to the chain of command, law enforcement (including Military Police or by dialing 911), or the Criminal Investigation Command (CID). The victim may also report an assault to Army Community Services (ACS), the Staff Judge Advocate (SJA), or by calling Army One Source at 1-800-464-8107 24 hours a day, seven days a week.

Sexual Assault in the Armed Forces

- *Number of sexual assaults by service members reported in 2006: 1,167*
- *Number of sexual assaults by service members reported in 2007: 1,516 (a 30 percent increase)*

All Soldiers must:

- *foster and promote a climate of mutual respect and interdependency among members of their units*
- *treat all other Soldiers with respect and dignity regardless of rank, position, sex, race, creed, color, or ethnic origin*
- *never knowingly place themselves or anyone else in a position of vulnerability to a sexual assault*
- *deal with all reports of attempted or accomplished sexual assaults in accordance with military regulations and procedures.*

Resources on Sexual Assault

The Army's sexual assault website, *http://www.sexualassault.army.mil,* has a wide array of tools and resources for you as an Army leader. The site has links to:

- The commander's checklist
- Evidence-collection procedures
- Printable documents and handouts
- Policy and regulations
- Confidentiality and restrictive reporting procedures.

Case Studies

Now that you are more familiar with what constitutes sexual assault, consider the following additional case studies from the *Consideration of Others Handbook.* Then take time to think about how you would answer the questions that follow.

Situation A: One night SPC Kendall went to the post theater with several other Soldiers. Following the movie, she said goodbye to her friends and took a shortcut through an unlit wooded area and a parking lot to get back to the barracks, which was nearby. As she walked across the lot, a stranger approached her. After attempting unsuccessfully to make conversation with SPC Kendall, he asked her if she was interested in having sex. SPC Kendall said no, but the stranger grabbed SPC Kendall and tried to kiss and fondle her. She pushed the stranger away and ran back to her barracks.

> *Discussion questions:* Did a sexual assault against SPC Kendall occur? What should SPC Kendall do now? What are some preventive measures SPC Kendall could have taken to avoid this situation?

Situation B: SFC O'Connor was the platoon sergeant for Company A. He was well respected by everyone in his unit. He visited his Soldiers on a regular basis in the barracks. On some weekends, he brought alcohol to the barracks to some of his Soldiers in the day room. PFC Scott had been in his platoon for three weeks; she normally stayed in her room whenever the Soldiers were down in the day room. One night SFC O'Connor knocked on PFC Scott's door and invited her to join the party. Although SFC O'Connor had had only two beers throughout the night, he brought PFC Scott seven or eight, and she became so intoxicated that she was barely able to walk or communicate. SFC O'Connor coaxed PFC Scott to lean on him while he helped her back into her room. PFC Scott passed out and later awakened to find SFC O'Connor having sexual intercourse with her. She was too intoxicated to resist or to remember the details of the incident.

> *Discussion questions:* Did SFC O'Connor commit a sexual assault? What are the effects on unit teamwork, morale, and readiness when someone in the chain of command sexually assaults a Soldier? What should PFC Scott do now? How can alcohol use contribute to the risk of sexual assault?

Situation C: PFC Dalton and PFC Williams were in the same unit and were roommates and friends. PFC Dalton had a reputation as a "ladies' man" and was attracted to a local woman he had recently met at a bar. Although they had not dated and had spoken only a few times, PFC Dalton felt that the woman might be attracted to him. PFC Dalton and PFC Williams were hosting a party at their apartment after they returned from a field training exercise, and the woman had already accepted their invitation. While planning the party, PFC Dalton told PFC Williams that this party was his chance to "make his move" on the woman. He bragged that after he made her one of his "special drinks," she would be "putty in my hands" and would "wake up next to me in bed." PFC Williams suspected that PFC Dalton's "special drink" contained a date rape drug

> *Discussion questions:* What are date rape drugs and how do perpetrators of sexual assault use them? What are PFC Williams' duties and responsibilities in this situation? What are the possible consequences if PFC Williams does not report this conversation and PFC Dalton's plan to the unit's chain of command?

Situation D: SFC Marcos had been serving as the Detachment Sergeant in Korea for more than seven months. He had earned a reputation as a hard working and caring leader. On the weekends, he checked on his junior enlisted Soldiers, especially the new arrivals. He was a likable person and many of his Soldiers, males and females, would welcome his visits. They would ask him to come by their barracks room to watch a movie, help prepare them for an upcoming board, or just talk. This was SFC Marcos's second unaccompanied tour in Korea and he remembered what it was like to be so far away from home as a younger Soldier. SFC Marcos befriended SFC Rhodes, and sometimes they would go to a club together. One night, SFC Rhodes invited SFC Marcos back to her room for a drink. SFC Marcos and SFC Rhodes had sexual intercourse later that evening. A junior enlisted Soldier saw SFC Marcos leaving SFC Rhodes's room early the next morning, and rumors began to spread within the unit about their affair. Two weeks later, SFC Rhodes was asked about the rumors by her commanding officer. Embarrassed by the situation, she accused SFC Marcos of rape. The CID began an investigation, but several weeks later SFC Rhodes recanted her story and acknowledged that the intercourse was consensual.

> *Discussion questions:* How might the Soldiers in his unit treat SFC Marcos differently after this incident? Could SFC Marcos have done anything to prevent this from happening?

Situation E: PFC Anthony was working in a deployed environment. After pulling a long shift of guard duty, he went to sleep in his tent. He later awakened to find a shirtless man on top of him in his cot. When he pushed the man away, the stranger ran from the tent.

> *Discussion questions:* What special sexual assault risks are present in a deployed environment? What steps can you and your unit take to reduce the risk of sexual assault in a deployed environment? What should PFC Anthony do now? What special issues do men face as victims of sexual assault?

CONCLUSION

The Army is built on unit cohesion, teamwork, and discipline to accomplish its larger mission of protecting the country, its people, and its values. To work productively and effectively, Soldiers need to feel secure, valued, and respected in their daily lives. The Army provides you with a comprehensive system of resources and training support to promote equal opportunity and prevent incidences of sexual harassment and sexual assault.

As a unit leader, you must educate your Soldiers about these issues, so they understand how their daily actions, attitudes, and words affect the other Soldiers in your unit. Ignoring these issues will have an adverse impact on the unit mission, cohesion, health, safety, discipline, and readiness. It will also negatively affect your combat effectiveness.

As the leader, you are responsible for the climate in your unit. You need to know if your Soldiers are experiencing—or perpetrating—incidents of sexual harassment or sexual assault so appropriate actions can be taken. In addition to the devastating toll they take on the victim, both cause your mission and your unit to suffer. You must do all you can to prevent such incidents.

Key Word

sexual assault

Learning Assessment

1. Explain the difference between the two types of EO complaints.
2. Define sexual harassment and explain why it is harmful to a military unit.
3. Define sexual assault.
4. Explain some ways to reduce the risk of sexual assault in your unit.
5. List steps to take when someone reports a sexual assault to you.

References

AR 27-10, *Military Justice.* 16 November 2005.

AR 600-20, *Army Command Policy.* 13 June 2002.

Consideration of Others Handbook. (n.d.). US Army. Retrieved 13 December 2005 from http://www.sexualassault.army.mil/files/Consideration_of_Others_Handbook.doc

Field Manual 6-22, *Army Leadership: Competent, Confident, and Agile.* 12 October 2006.

George Mason University Sexual Assault Services. (2005). Retrieved 13 December 2005 from http://www.gmu.edu/facstaff/sexual/

US Army Sexual Assault Prevention & Response Program. (2005). US Army. Retrieved 13 December 2005 from http://www.sexualassault.army.mil

ARMY CUSTOMS AND COURTESIES

Key Points

1 Army Customs and Courtesies

2 Army Etiquette and Protocol

3 Some "Don'ts" in Army Culture

Courtesy among members of the Armed Forces is vital to maintain discipline. Military courtesy means good manners and politeness in dealing with other people.

Field Manual 7-21.13, *The Soldier's Guide*

Introduction

Army life is certainly not all hard work. Army customs and courtesies, particularly the social customs, instill pride, cohesion, and a sense of identification with the organization. As a soon-to-be officer headed toward your first assignment, you'll want to understand these customs and courtesies, both to avoid embarrassment and to quickly establish your credibility as an Army leader.

When you report to your first unit, you will find that in addition to Army customs, your unit members will have adopted or created additional customs as part of their specific unit culture. Recognizing and adopting these customs will enable you to win your Soldiers' confidence and trust.

> NOTE: Army custom requires you to report to your first duty station as a new lieutenant wearing your Army Class A uniform. New lieutenants who pack their Class A uniform in their carry-on bags and report to their new commanders wearing their Class As will be looked upon favorably. But new lieutenants who report to their new commanders in their Army combat uniform (ACU)—because the movers packed their Class As in their household goods to be shipped to their new duty stations—will be looked upon in a negative light.

Your success as a leader will depend in large part on your willingness to practice and uphold Army traditions and customs and to practice courtesy toward all military personnel. These traditions are the lifeblood of the Army family.

As a young lieutenant, Colin Powell learned something of the importance that little courtesies can have, not only for Soldiers, but for their families as well.

The "Stork-Alert" System

My tour as commander of A Company was short. [In 1961] I was sent off to become adjutant of a new unit. . . . Once again, I was a first lieutenant in a captain's job. A battalion adjutant handles personnel, promotions, assignments, discipline, mail, and "morale and welfare." My new commander was Lieutenant Colonel William C. Abernathy.

Lieutenant Colonel Abernathy was no swashbuckler, but he was a solid performer who gave troop morale top priority. He expected a promotion to private first class to be handled with the same importance as a promotion to colonel. The men were to be paid on time. Soldiers freezing their butts off in the field were to have hot coffee and soup available. Any sign that a GI was not being properly looked after meant trouble right up the chain of command. Abernathy did not pamper the troops; he worked them hard and disciplined them, which was another way of caring.

One day, the colonel informed me that I was to set up a system of "Welcome Baby" letters. My mystification must have shown in my face. Every soldier whose wife had a baby, Abernathy explained, was to receive a personal letter from the battalion commander congratulating the parents. A second letter would go to the baby, welcoming the tot into the battalion. Abernathy demanded that I get these letters out the very day the child was born.

How was I supposed to know which men were about to become fathers? I could picture the battalion, massed on the parade ground: "Every man whose wife is pregnant, take one step forward! All right, when's she due?" I suspect my bachelor status also had something to do with my lack of enthusiasm. In any case, I dragged my feet in setting up this stork-alert system. Abernathy called me on the carpet. "Gee whiz, Colin," he said. "I'm disappointed you haven't done this yet." I would rather have had [another officer] blister me with four-letter words than hear Abernathy's pained reprimand. I returned to my office and immediately added population reporting to my duties.

To my surprise, once we had the system in place, we started getting positive feedback. The soldiers were impressed by Abernathy's thoughtfulness. Mothers wrote us that they appreciated being considered part of their husband's Army life. The babies were not talking yet, but I imagine that, somewhere out there, a thirty-five-year-old woman is wondering how a letter making her a member of the 1st Battalion, 2d Infantry, got into her baby book.

Another lesson learned and filed. Find ways to reach down and touch everyone in a unit. Make individuals feel important and part of something larger than themselves. Abernathy had found a way to demonstrate caring in a fundamentally rough business.

GEN Colin Powell

Critical Thinking

There is a very old Army saying that if the Army wanted you to have a wife, it would have issued you one. What difference does it make if a spouse feels as if he or she is a part of Army life? After all, the Soldier, not the spouse, works for you.

Army Customs and Courtesies

All cultures consist of various forms of **customs** and **courtesies**. They foster civility and order in society. Within organizations, they create pride and cohesion. In the Army, routine encounters between Soldiers and officers using such customs and courtesies encourage respect and loyalty. In formation, they cultivate order, consistency, and esprit de corps.

Customs

Army customs conform to the code of the Army officer. They also conform to the established rules of military courtesy.

Many customs observed in the Army originated in antiquity. They are time-honored forms of military pageantry, most of which are followed, in one way or another, by armies around the world.

For the most part, customs consist of acts that honor the nation's flag or military dead or that pay respects to comrades-in-arms. These customs help forge the bonds of unit cohesion and pride. They are the basis for the perpetuation of consistent Army practices. They form the Army culture.

You are already familiar with a number of these customs, including the hand salute and raising your right hand when taking oaths. When you live on post or on station, you will become familiar with the customs of daily military life, such as the morning reveille, the evening gun salute during retreat, and taps. As your military career progresses, you will inevitably see many examples of military customs through ceremonies. Your unit will certainly participate in the elaborate but festive post-wide, 21-gun salute on Independence Day or on the Army's birthday. You may have the good fortune to observe or participate in the ceremonious but lighthearted saber arch over the bride and groom at military weddings. Or you may have the somber honor of hearing the three volleys by the seven-Soldier honor guard and the playing of taps to honor a fallen Soldier.

Courtesy

Courtesy likewise facilitates camaraderie and unit cohesion. Observing courtesy is essential in maintaining human relationships. It stimulates harmony among individuals, smoothes the conduct of daily interactions, and adds a welcome note to all manner of human contacts, civilian and military. Courtesy is critical to teamwork and to getting things done.

Prescribed courtesies are a part of the ceremonial procedures that add color and dignity to your military life. They contribute to the comradeship that binds—at all grades and ages—those who share a common responsibility for the nation's security.

As an officer, you are a role model sworn to uphold the values, customs, and courtesies of your unit, the Army, and the country. Learning and practicing Army courtesies is an important part of your development as a career officer.

custom

a common tradition or usage so long established that it has the force or validity of law

courtesy

polite behavior; a polite gesture or remark

Comradeship and Army Social Customs

Comradeship among your military associates is the strongest and most enduring of Army customs. Ask Soldiers who have served in the Army and many will happily tell you what this comradeship means to them. In the military profession, trusting one another implicitly is crucial—lives depend on it—and customs and courtesies help build that foundation of trust and fellowship.

One very strong Army custom is the *Officers' Call*. This is an informal gathering of unit officers at the local Officers' Club (O-Club) or other local establishment for drinks and conversation. The custom establishes a bond between officers and fosters the sharing of knowledge.

Note:

Most officers' clubs will have a bell at the bar to guard another time-honored tradition. The ringing of the bell notifies everyone at the bar that someone has placed his or her headgear on the bar counter—a lighthearted infraction requiring the offender to buy a round of drinks for everyone at the bar.

Units may hold *social functions* to celebrate holidays or promotions, to commemorate events in the units' history, or to honor the occasion of arrivals and departures. Holiday social functions may include the unit Christmas party—which may be catered or potluck—at which families get together to swap gifts or children visit the unit's Santa, who is normally played by one of the jollier Soldiers in your unit. For Thanksgiving, your Soldiers may be invited to wear their dress blues to attend the Thanksgiving meal with their families held at the unit dining facility (DFAC). (Officers always wear their dress blues to this meal.) Other unit social functions may include unit barbecues to celebrate Independence Day or luncheons to honor POWs and MIAs from past wars.

Most units have a long and prestigious history and plan social functions to honor the unit's lineage. Many units will plan socials to celebrate the Army's birthday, the unit's success during certain war campaigns, or the unit's notoriety at critical battles. Your unit may be invited to participate in an international social function, such as the annual Nijmegen March—a 120-mile road march completed in 30-mile legs over a 4-day period and centered on the historical town of Nijmegen, the Netherlands. Army units that participate in the Nijmegen March will have friendly competition with army units attending from around the world and will no doubt be a part of many international social functions before, during, and after the famous march.

Most battalions will have a monthly social for their officers and senior noncommissioned officers (NCOs) (first sergeants and above) called the battalion hail and farewell. During the monthly hail and farewell, the battalion commander will "hail" all new officers and senior NCOs and their spouses by introducing them to the battalion's leadership and telling a little about the officers' and wives' backgrounds. The battalion commander will also say farewell to all who are about to depart the unit either through permanent change of station (PCS), expiration term of service (ETS), or retirement. During the farewell, the battalion commander may present to departing Soldiers their PCS, ETS, or retirement awards and other recognition or gifts—usually in the form of a unit-specific plaque or other memento. The commander normally recognizes a departing Soldier's spouse for her or his service as an Army spouse, usually in the form of a unit-specific gift, plaque, or flowers.

In addition to the battalion hail and farewell, Soldiers and spouses may also be hailed or farewelled at all levels of the command, from the squad or platoon up through the company. If you are assigned to the battalion staff, you may have a staff-level hail and farewell as well. In addition, the family support groups at the platoon and company level may also conduct similar hail and farewells for the Army spouse.

Sometimes units conduct social functions just for the fun of it or as a method of fostering esprit de corps or camaraderie and reducing stress. Many times the operational tempo (OPTEMP), frequent deployments, or manpower turnover can produce anxiety and stress within the officer corps. Good Army leaders will recognize this and plan impromptu parties. These events have one simple rule: You cannot discuss work.

Some units may also have a monthly or quarterly social that is entertainment driven. Planning or hosting the social rotates to each company or staff section. Units get a lot of fun out of planning an event that will outdo the last social, with the emphasis on doing something that nobody has planned before. The hosting unit tries to keep the plans a secret until the day of the event to build curiosity and excitement. The host unit has the latitude to be creative and usually receives guidance or limitations from the battalion or company commander—such as cost of the event, time it takes to complete the event, or distance required to drive to and from the event. Some units plan fun socials such as bowling, paintball, shopping trips, guided tours, or simply dining out at a favorite local restaurant. At the event's conclusion, the commander will usually, in seriousness or in jest, compliment or roast the event planner on a job well done. This, of course, raises the bar for the leader who must plan the next social function.

A *Dining In* is a formal dinner function for members of a military organization or unit. It requires detailed planning, coordination, and execution. As a second lieutenant, you may be detailed to help the battalion adjutant organize a Dining In. A *Dining Out* is a variation that includes spouses. Dining Ins and Outs are considered a *command performance* and as an officer, you should plan to attend these formal dinners. Declining to attend requires a justification to your company commander, who will pass this information up to the battalion commander. Not attending should be the exception rather than the rule—reasons for not attending would include such events as attending a family wedding, a funeral, a pre-approved leave, or a serious illness in your immediate family. As an officer, you are expected to manage your funds wisely, so "being broke" should never be a reason for not attending a unit function. Doing so would reflect poorly on your responsibility as an Army officer.

Command performance means that your superior officers expect you to attend an event.

Special Events

As you will no doubt discover, if you haven't already, the Army is a very social family. Army officers celebrate many life events at a gathering of some kind, just as many everyday American families do. Some special events include:

Military weddings. Additional customs may include forming the saber arch, using a saber to cut the cake, and displaying the national and unit colors. All officers should wear an appropriate military uniform, such as dress blues. An unofficial but highly practiced custom at Army weddings is for the last Soldier in the saber arch (normally the groom's best man) to gently swat the bride on the backside with his saber as she passes from under the arch and loudly proclaim "Welcome to the Army." This custom is similar to the civilian practice of removing the bride's garter or throwing the bouquet, and those who know what is about to happen to the unsuspecting bride look forward to it.

Christmas at the White House

Birth of a child. Your unit commander may send a personal letter of congratulations to the parents on behalf of the entire command, and many units will mark the arrival of the newest member of the command with an engraved silver cup and flowers for the new mother. Most commands have a "Cup and Flower Fund" to which every officer makes a monthly donation. The battalion uses the fund to purchase cups and flowers for new arrivals as well as send flowers on the command's behalf to a Soldier's family who has lost an immediate family member. Most units will "hail" the newborn child at their next hail and farewell.

Unit picnics, cookouts, and parties. The most important unit social function you will attend is the unit party. Some units plan this social in the summer and may refer to it as their unit picnic. Others may hold their unit party during the Christmas season. The unit party is partly funded by Morale, Welfare, and Recreation (MWR) funds. Each command receives an MWR fund based on the number of Soldiers assigned to the command. The unit party is scheduled on the unit's training schedule and therefore is a mandatory training event that every Soldier present for duty must attend.

In addition to the unit party, other units at various levels arrange informal get-togethers or cookouts throughout the year for morale, team building, and esprit de corps. As a lieutenant, you may also be invited to attend informal socials planned and attended by other second lieutenants or platoon leaders.

AR 215-1 governs unit MWR fund accounts and unit fundraising. If you are tasked to plan, coordinate, or execute any part of a unit activity involving unit funds, be certain that you read and understand the regulations. Failure to do so can result in charges of misappropriation of government funds.

Your participation at events hosted by your company or battalion commander is usually mandatory. Courtesy implies that you respect and honor the occasion with your presence. It's not permissible or mannerly for you to just "make an appearance." Rather, you are expected to remain for the duration of the event. If you must depart early, military courtesy is that you inform your commander of the requirement prior to departing and explain your early departure (such as the babysitter must be home by a certain time, or your spouse has to be at work early the next morning). It is always best to explain these circumstances prior to the unit social so that your commander is informed ahead of time.

There's no requirement to stay at any function until the last person leaves. The host will generally make an announcement ending the official festivities and invite those who wish to remain longer to do so. Most people who need to leave early will begin to depart with this announcement. In fact, it's a point of good manners not to be the last to leave a function or to try to "close down the bar." Use your good judgment and discretion in such matters.

Note:

The Army culture and attitude on drinking has changed dramatically over the past decade, with an emphasis on responsible drinking and zero tolerance of driving while impaired. In the recent past, it was customary and acceptable behavior in many units to drink yourself under the table. But it is no longer acceptable in today's Army culture to get "hammered" at social functions, even if held at a bar or a private residence. As a newly commissioned officer, you must leave whatever bad drinking or partying habits you may have picked up on campus. The Army careers of many second lieutenants have been ruined within the first few months by irresponsible drinking at their very first unit social function. Drunken behavior at your new unit is a stigma that you cannot overcome, and it will most likely be reflected in your officer evaluation report (OER).

Critical Thinking

What importance does the camaraderie built during social events have during combat or other operations? Is there ever a downside to this camaraderie?

Customs in Connection With Sickness and Death

Army officers feel a special responsibility to their Soldiers, their fellow officers, and the families of both. When Soldiers are sick or die, officers are quick to express their concern or condolences. It is even more the case when a Soldier's spouse or child becomes seriously ill or passes away—this is an emotional event that will bring the spirit of family, camaraderie, and unit cohesion together in order to provide love and support to that Soldier and his or her family. As mentioned earlier, the unit will use its Cup and Flower Fund to provide a small gift of love and support in that Soldier's time of need. In addition to the unit's Cup and Flower Fund, it is both custom and courtesy for the Soldier's leadership to visit the family, attend the viewing and funeral services if geographically possible, and offer a personal gift.

Visiting the Sick

As an officer, you are expected to visit Soldiers in your unit who are ill as the situation allows. If you are married, your spouse is also expected to visit the spouse of a hospitalized Soldier and offer support. As a general rule, if it is custom or courtesy for you to offer a gift or make a personal appearance, your spouse will also be expected to do the same— that is, to offer a gift from your family or make visits with you when a member of a Soldier's family, such as a spouse or a child, is ill.

Death of a Soldier

When a Soldier dies, the Army assigns a Casualty Assistance Officer (CAO) to the case. The CAO is responsible for assisting the next of kin in all matters relating to the deceased Soldier's pay, benefits, and belongings. The CAO acts as the conduit between the next of kin (who may be unfamiliar with Army policy and procedures), the command, and the installation support staff, who will assist in returning the deceased's property and remains to the next of kin.

Many officers will go through their entire career and never receive a CAO tasking. The task of informing and interacting with the next of kin who have lost their loved one is a very difficult but honorable and important task. If you are asked to serve as a CAO, it will take a great deal of intense fortitude to maintain your military bearing—but you must display confidence and strength in order to help the family through this difficult time.

If you are a leader in the deceased's command, you will be expected to attend the unit's funeral service. If it's geographically feasible, you should make every effort to attend the private viewing and services held by the deceased's family. Having a large contingent of Soldiers show up for the funeral will leave a positive impression not only with the deceased's family members, but also with everyone who attends the services.

The unit will send flowers purchased with the unit's Cup and Flower Fund to the funeral home at which the services are held and a sympathy card to the next of kin's home. Unit officers and their spouses may also send flowers themselves or on behalf of the unit. The unit commander writes a letter of condolence on behalf of the brigade, battalion, or company. If you are assigned as the battalion's assistant administrative officer (S-1), you may be required to proofread the commander's condolence letter before mailing it.

Support of Post and Organization Activities

As an officer, you should plan to support the activities of the unit and garrison to which you are assigned. It's customary for officers and their spouses to attend social events sponsored by enlisted members when you are invited. Remember that your behavior is the standard, so drink and behave in moderation. It should go without saying that excessive drinking and unruly behavior are *never* appropriate for an officer. Depart with or immediately after the senior officer—never remain at an enlisted social function without your platoon sergeant.

Likewise, it is inappropriate and not in keeping with military customs for an officer to invite one or more enlisted or junior NCOs over to his or her home for a formal or informal social function without first discussing it with the platoon sergeant. If he or she recommends the social, you should ensure that your platoon sergeant also attends.

There are some exceptions to this rule. For example, it is customary for leaders at all levels to survey their Soldiers about who has plans for traditional holidays centered on meals, such as Thanksgiving, Christmas, and Easter. For single Soldiers who are geographically separated from their families during these holidays, or for a Soldier and his or her family who have nothing planned for these holidays, it is customary for the Soldier's leaders to invite the Soldier or Soldier and his or her family over for the holiday meal. This is especially true for units outside the continental United States (OCONUS), where it is often financially unfeasible for Soldiers or their families to travel home. Regardless of the holiday or event, you must be especially cognizant of your conversation and behavior to ensure you do not cross the lines of professionalism among officers, NCOs, and the enlisted ranks. You do not want to give your guest or others in the unit the perception of favoritism or create the perception that fraternization took place.

You should also schedule time to attend athletic or community events in which members of your unit are participating, such as post intramurals or tournaments, including softball, basketball, flag football, boxing tournaments, marathons, and triathlons. Many commanders encourage their officers to participate in or coach such teams. Team and individual success can often be captured on your OER support form as achievements that can favorably affect your OER.

Critical Thinking

Compare and contrast Army social functions with team-building activities.

Newly commissioned officer rendering his first salute

Army Etiquette and Protocol

Proper **etiquette** is not cursory or superficial; it is a practical set of rules that will help guide you and keep you out of trouble during social functions. When you learn them, these rules save time that you would waste in trying to decide what is proper. Etiquette helps people proceed with the more important phases of social interaction.

Familiarize yourself with military etiquette and the correct way to address the various titles and ranks of people you will encounter in ordinary Army life. Address commissioned officers as Sir or Ma'am—not by rank—and warrant officers as Mr. or Ms. You should address NCOs by rank and Cadets as "Cadet" or as Mr. or Ms. Address Army retired personnel as Sir or Ma'am and civilians as Sir or Ma'am or Mr., Mrs., or Ms.

If hosting a social function at your home, it is important for you and your spouse to understand seating etiquette; that is, how to seat your guests based on rank and gender.

Customs of Rank

Besides saluting, there are several other ways you recognize and show respect for an officer's higher rank. The "place of honor" is on the right. When walking next to a higher-ranking officer, you will walk to his or her left. At various functions, the guest of honor sits or stands on the host's right. When departing a reception or social event, junior officers should depart after the commanding officer. Although not required by regulation, it is both customary and courteous to salute and greet senior officers when off duty and in civilian clothes, if you recognize them as senior officers. Likewise, protocol requires you to salute colonels or general officers in their vehicles if the vehicle displays their rank on the vehicle placard. It is only prudent, however, to render the appropriate salute if you indeed recognize them, can clearly identify their rank, or see their rank decal in their window.

Courtesy Calls

By tradition, officers arriving at their new duty station make brief courtesy calls to superior officers in their chains of command. This enables your superior officers to meet you and connect your face with your name soon after you arrive. You should ask your unit adjutant which officers you should call on, and you should leave a business or calling card during your visit.

New Year's Commander's Call

Army tradition holds that officers and their spouses make a formal call on the commanding officer during the afternoon of New Year's Day. Particulars vary depending upon station and current events—occasionally the commander designates an alternate day. The commander might send invitations or will otherwise let you know the date, time, and location of the event. Officers usually wear their dress blues, but the commander may dictate the uniform. It is also appropriate to leave a business or calling card during the visit.

Appointments With the Commanding Officer (CO)

When you wish to meet with a battalion commander or higher, you should make an appointment. You do this through the battalion adjutant, executive officer, or an aide.

Permission of the First Sergeant

It's customary for a Soldier who wishes to see the company commander to request permission from the company first sergeant. As a platoon leader, you should certainly be aware that one of your Soldiers has asked to visit the commander, and the first sergeant will notify you.

Open Door Policy

The open door policy permits everyone in the Army, regardless of rank, to appeal to the next higher commander. As an officer, you should expect this and not take offense. At the same time, a Soldier should make every attempt to resolve a problem using the chain of command before going to the next higher command.

Local Customs and Traditions

You should become acquainted with local traditions and customs at each new duty station. It's important to know your unit history and its particular customs and traditions. Ask your commander about special company and battalion ceremonies and functions, such as the Grog Bowl. Talk to your platoon sergeant and company first sergeant about particulars of the unit's customs; they usually can provide a wealth of history and tradition. When you are posted to a foreign country, make a sincere effort to learn about local customs, courtesies, and events. Avoiding cultural missteps can contribute significantly to the success of your mission.

Note:

The Grog Bowl ceremony is a unit ceremony in which the history of the unit is recited using a recipe of spirits or other unique ingredients that are poured and combined into a common bowl. Soldiers in the unit take a common drink from the bowl to show their allegiance to the unit. An ingredient is added for each major or significant historical campaign or event that the unit was involved with. As the narrator recites the unit history, he or she gives a brief explanation of how the ingredient ties in to the unit colors or history. Once the Grog Bowl ingredients are complete, the unit commander selects a Soldier to test the fitness and suitability of the drink. After the unit commander approves of the Grog, each Soldier takes turn in dipping his or her cup into the Grog Bowl, and toasts are given to the Army, the Regiment, and on down to the battalion.

Military Funerals

As an officer, you may also at some point either attend a military funeral or act as a Casualty Assistance Officer. You should know the proper ceremonies and procedures for a military funeral, as described in AR 600-25 and FM 3-21.5. In particular, be familiar with:

- the significance of the military funeral
- courtesies at a military funeral
- the badge of military mourning
- the elements of a military funeral ceremony (normally selected from a rotating special duty roster [DA Form 6] by the battalion adjutant or the installation or post adjutant general)
  - honorary pallbearers
  - the firing party
  - the bugler
  - those charged with attention to families
- appropriate attire
- the right words
- introductions.

Army Social Customs

As a junior officer, you usually aren't expected to do a great deal of formal entertaining. You should conduct some platoon social activities, however, to help build a good team and to get to know your Soldiers. Some other customs you should be aware of include:

Officers' Club (O-Club). Once assigned to a permanent station, your fellow officers will encourage you to become a member of the Officers' Club if there is an active one on post. It's worth trying it out to learn the benefits of membership.

Cultural opportunities. Part of the role modeling you will project to your Soldiers should involve taking part in the many cultural opportunities offered by Army life. Cultural events and community activities can add color, depth, and character to life. Attending such events will help you understand and lead your Soldiers. Your installation Equal Opportunity (EO) office and the civil service Equal Employment Opportunity (EEO) office also sponsor cultural events. These events recognize and honor the significant accomplishments of women, African Americans, Native Americans, Pacific Islanders, and so on. Most events will offer a chance to observe cultural displays and native attire as well as partake in ethnic food and drink.

Community service. Many organizations on post welcome volunteers. These outreach activities usually benefit both the community and the military station. Members of Army families are always encouraged to participate. You and your Soldiers will have many opportunities to volunteer to coach sports teams for children ranging in age from 4 to 18 at your installation Child and Youth Services office. You will also have many opportunities to volunteer for local churches and other organizations.

Social and recreational opportunities for dependents. Most posts have a range of thriving recreational organizations for officers' and Soldiers' children, including league sports, clubs, associations, Scouting, and religious activities. The parents of children who participate are also encouraged to get involved in these activities. Each installation will have an MWR office that will provide you with information on low-cost or free recreational opportunities available only to Soldiers and their family members. There are hundreds of Army MWR resort areas around the world where Soldiers and their families can rest and relax at reduced rates. These include such places as Hawaii, Korea, Germany, Disney World, and elsewhere throughout the United States. In addition to enjoying Army MWR resorts, you and your family can also vacation at other Defense Department (DoD) MWR facilities run by the Navy, Air Force, and Coast Guard, as well as visit locations throughout Europe—such as Spain, Iceland, Greece, and Italy— and throughout the Pacific and Asia—such as Singapore or Guam. Army and DoD MWR offices also offer discounted vacation and tour packages, as well as discounted tickets to theme parks and cultural and recreational events across the globe.

Building Social Goodwill

By now you have learned that team building and networking are important leadership skills. As a leader, you should try to be on good terms with everyone. Basic etiquette and politeness help you achieve this. When you receive a social invitation, express appreciation at being included. Pay particular attention to written invitations and *always RSVP* (that is, respond accepting or declining the invitation) on time. It is important to know that the invitation is extended only to those addressed on the invitation. You should pay particular attention to how the invitation is addressed so that you are not professionally embarrassed by bringing someone to the social event who was not included in the written invitation. For example, if your name is on the invitation but your spouse's or date's name is not, that means it would be inappropriate to bring your spouse or date. For any social affair, arrive promptly. If appropriate for the occasion, you should bring a hospitality gift such as a bottle of wine or flowers. Remember to send a note of thanks afterward. Most important, you must RSVP no later than the date specified to indicate whether or not you will attend and how many invited guests you will be bringing. If you cannot attend, simply RSVP "with regrets."

The Right Clothes

Make sure your choice of clothing fits the occasion. If you're unsure of the dress code for an event or occasion and can't get clarification, err on the side of dressing up. Know and use the standard Army dress when possible. For example, with your dress blues, you usually wear a bow tie after 6 p.m. and a four-in-hand tie before 6 p.m.

Often an invitation to a social event will specify the dress for the occasion. Here are the most common terms and the type of clothing they refer to:

Formal—tuxedo for men, evening gown for women; mess dress or dress blues

Informal—business suit for men, cocktail attire for women; dress blues

Casual—slacks with a sport shirt or jacket for men, a skirt or dress for women

Athletic attire—clean, well-kept sportswear.

Note that *informal* and *casual* are not the same thing!

The Right Words

When you attend a social event or formal occasion, remember to mingle. Visit with and speak to as many of the other guests as possible, and try to learn their names. Have business or calling cards that you can give those you meet to help them remember your name. Be a good listener and ask about their families, places of origin, and hobbies. Don't monopolize the conversation and avoid "shop talk" that excludes people from the conversation. Stay away from controversial topics—such as politics and religion—and avoid off-color jokes. Make sure you know and understand the correct procedure for making introductions— a very important social skill.

Military Protocol

Understanding basic military **protocol** is essential. A good reference for this is DA PAM 600-60. While this pamphlet covers much of Army and social protocol, you should learn as much as you can from the activities you attend. What you learn now will be very helpful later. For example, as a platoon leader you will host few ceremonies and events. As a company commander, however, you will be much more involved in these types of activities. Become familiar with the requirements before you assume the role. Events that will test your knowledge of protocol are:

protocol

a code of correct conduct

- Formal receptions and receiving lines: Know how to go through a receiving line (Never with a drink, a cigarette, or a cigar!)
- Attending any social activity: Be familiar with what to wear, what to bring, and how to introduce your spouse or date
- Acting as a host or a hostess: Know how to arrange and conduct a receiving line, proper table seating, presentation order, and introductions.

Some "Don'ts" in Army Culture

As a leader, you must realize you set an example to your Soldiers in everything you do. They will talk about you and your actions and may repeat them. While no Soldier should display poor social qualities, you attract extra scrutiny as a leader.

Some actions you should be particularly careful about:

fraternize

improper socializing between an officer and an enlisted Soldier; also, to associate on close terms with members of a hostile group, especially when contrary to military orders

- Don't defame the uniform by wearing unauthorized items. Do not wear the uniform where it is not appropriate
- Offer no excuses, even if you think you have a good one
- Don't improperly **fraternize** with subordinates or enlisted Soldiers
- Don't be servile. Don't fawn over the commander. This doesn't mean you should avoid the commander, however—just be professional
- Avoid praising the CO to his or her face
- Never lean on a superior officer's desk or act too familiar with your commander
- Avoid going over an officer's head when possible. Let your commander know when you must visit his or her boss using the open door policy
- Never keep anyone waiting—arrive early
- Avoid harsh remarks—don't participate in or encourage gossip, slander, faultfinding, and so on. This will erode your leadership and destroy unit trust and cohesion
- Avoid vulgarity and profanity—it's unprofessional and inappropriate, and offending someone will damage your leadership ability
- During the retreat ceremony, don't slink back into the office and pretend you didn't hear the signal
- Don't carry an umbrella while you are in uniform—it's against regulations.

Note:

On most installations, it is expected that when you hear reveille or retreat you will safely pull your vehicle over and exit your vehicle in order to properly participate in morning reveille (salute until the last note of reveille) or to properly participate in retreat (parade rest for "to the colors," then salute and hold salute until the final note of retreat). Don't keep driving and pretend not to hear the music—as someone who knows you and your commander will surely see you. You must plan extra driving time to account for being "caught" by reveille or retreat to avoid being late for appointments or formations.

Critical Thinking

Why would servility toward a superior officer or fawning over the commander be a problem for you? How might it affect your Soldiers or your fellow lieutenants, whom this commander also rates, who witnessed this kind of behavior?

Modeling Moral Standards

Your Soldiers, peers, and chain of command expect you to demonstrate the highest moral fiber as an Army officer. You are on stage 24 hours a day. Your actions—or inaction— speak volumes about your character. A display of poor character could tarnish your reputation for the rest of your career, if not your life. The old adage that "silence implies consent" is a valid perception when it comes to remaining silent when present or around others whose behavior may be questionable or not in accordance with Army Values.

The camaraderie of Soldiers requires trust and credibility. After all, you will entrust your life to your fellow Soldiers and they will entrust theirs to you. Would you want to follow someone of questionable character?

Critical Thinking

What would you do or say if you were present in a group of fellow lieutenants when one of the lieutenants made an off-color joke or comment about race or gender? Would you remain silent since you did not make the joke or comment? Would you laugh with everyone else? Would you address this as a breach of Army Values? If so, would you make an on-the-spot correction, or would you wait and address the lieutenant privately or off to the side? How will your response affect trust between you and your fellow lieutenants in the future? Is it more important to enforce the Army's values with your peers or to hold their personal trust?

Even so, at some point you will have to deal with others of questionable character. In doing so, you must always keep the Army Values in mind. If that person is a Soldier, correct and discipline the individual. If it's a peer, you may need to distance yourself from that person or discuss the offensive behavior with him or her directly. If the behavior does not change after you speak with her or him, you may have to address the behavior with your commander. If it is a superior, you must act. Never accept the notion that you are compelled to do something morally or ethically wrong—or to do nothing in the face of wrong. After careful thought, decide how best to handle the superior officer, either through direct conversation or through the proper military channels. When addressing a superior officer's morals or values, it is always best to do so in the presence of another trusted officer in order to avoid any perception, characterization, or claims that you were disrespectful when addressing your concerns with the superior officer.

CONCLUSION

The proud exercise of military customs and courtesies is a time-honored foundation of Army life. In your unit, such observation is a clear sign of your Soldiers' morale—and of your good leadership. Such observance also demonstrates your organization's ability to perform under stress as a team of professionals bound by the Warrior Ethos and by their mutual respect.

By embracing military customs and courtesies, you lead by displaying appropriate military behavior and values to your Soldiers. You demonstrate that you care about your Soldiers and their families as well as your fellow officers and their families.

Army personnel and families share special challenges and difficulties that their civilian counterparts rarely face. The courtesies military people exhibit are an important way they help one another meet those challenges and overcome those difficulties.

Key Words

custom
courtesy
etiquette
protocol
fraternize

Learning Assessment

1. Explain how Army customs and courtesies build unit cohesion and pride in the organization.

2. Describe some of the Army social functions you will attend as an officer.

3. Explain some of the etiquette and behavior that builds social goodwill.

4. List some of the social behavior you should avoid as an Army officer.

References

AR 600-25, *Salutes, Honors, and Visits of Courtesy.* 24 September 2004.

Bonn, K. E. (2002). *Army Officer's Guide.* 49th Edition. Mechanicsburg, PA: Stackpole Books.

Field Manual 3-21.5, *Drill and Ceremonies.* 7 July 2003.

Field Manual 7-21.13, *The Soldier's Guide.* 2 February 2004.

PAM 600-60, *A Guide to Protocol and Etiquette for Official Entertainment.* 11 December 2001.

Powell, C. (with Persico, J.). (1995). *My American Journey.* New York: Random House, Inc.

INTRODUCTION TO BATTLE ANALYSIS

Key Points

1 **Battle Analysis Checklist**

2 **The Staff Ride**

3 **Abbreviated Case Study:**
The Battle of Kasserine Pass

Training and leader development must include
a historical perspective—especially of the conduct
of battle.

GEN Frederick M. Franks Jr.

Introduction

As an MSL II Cadet, you were introduced to the principles of war. As an MSL III Cadet, you applied those principles to the Civil War Battle of Chancellorsville. The method you used in that case study can be described as an abbreviated application of **battle analysis** that focused on the principles of war. The US Army Command and General Staff College developed the battle analysis method to help students structure their studies of battles and campaigns. Any military professional seeking insight from historical battles and campaigns can easily apply the format to help deepen his or her understanding of warfare and the profession of arms.

This section will review the battle analysis method and then present a partial case study of the Battle of Kasserine Pass in Tunisia during World War II.

battle analysis

a method used by the Army to provide a systematic approach to the study of battles, campaigns, and other operations—it is designed as a general guide to ensure that significant actions or factors affecting the outcome of a battle or operation are not overlooked

Soldiers engaged in battle in World War II

Battle Analysis Checklist

The battle analysis process is a checklist that ensures that you examine all the critical aspects of the battle or the campaign in question. The checklist is divided into four sections:

- Define the subject
- Set the stage (strategic, operational, and tactical settings)
- Describe the action
- Draw lessons learned.

First you decide which battle to study. Once you've chosen a battle, you gather the information necessary for a thorough and balanced study and organize it logically so you can analyze it. Then you perform the analysis and list the lessons learned.

You don't have to follow the checklist to the letter, or even use every part of it in your study. Don't let the format's order disrupt the flow of your study, either—but be sure at least to consider all the elements.

Define the Subject

Like any military operation, your study of military history needs a clear, obtainable objective. So you should begin by defining what you will study. Determine what, where, when, who, and why. Frame your study by determining the date of the battle or campaign, its location, and the adversaries involved.

Next look for good sources that will help you make a systematic and balanced study. You can use books, articles, the Internet, video, audio, and other electronic means.

Look for a variety of books to get a balanced account of the battle. You should consult memoirs, biographies, operational histories, and institutional histories for information on your subject. Don't overlook general histories, which can help provide the battle's strategic setting.

Articles from professional military publications and historical journals can also be excellent sources of information. Video and film documentaries containing footage of actual events or interviews with people who took part in a battle can add to your understanding of the events. Check to see if transcribed oral history interviews with battle participants are available. In addition, check the Internet for electronic documents on more recent military operations as well as historical campaigns.

Regardless of what you're researching, it's always useful to evaluate your sources. Despite the large volume of published material and the enormous amount of raw information available on the Internet, finding good sources is not always easy. As you gather your research material, you should consider each potential source in terms of its content and bias. What information can the source give you? Is it relevant? Will it help you complete the study? Is there a clear bias, and if so, what is it? Does the bias interfere with your use of the source? Some sources are so biased that their credibility is suspect. However, a source can be extremely biased yet still contain useful information or observations.

Set the Stage

You need a good understanding of the strategic, operational, and tactical situations before you can analyze the battle. The amount of detail you go into depends on the purpose of your study and the audience you're addressing. If everyone knows the causes of the war and the opponents, for example, you may not need to describe these in much detail. A few paragraphs may be enough to place the battle in its proper context. For example, you probably don't need an abundance of detail on the causes of World War II to analyze the Battle of Okinawa. But you will need a thorough knowledge of the campaign in the Pacific.

First you should consider the *strategic factors:* What caused the war? Who were the opponents? What were their war aims? What armed forces did the adversaries possess? How well trained, equipped, and armed were they? Which significant social, political, economic, or religious factors influenced the armies?

Next describe the *operational settings:* What campaign was the battle part of? What were the campaign's objectives? Did any military factors—alliances, tactics, doctrine, or personality traits—affect the campaign? How did the battle fit into the overall campaign?

Then review the *tactical situation:* Since these factors have a direct effect on the operation, this part of the format will often answer why a commander took or didn't take a particular action. You study the area of operations much as you have learned to do as a platoon leader. What was the weather like in the area of operations? How did it affect the operation? Use OAKOC (observation and fields of fire, avenues of approach, key and decisive terrain, obstacles, cover and concealment) factors to describe the terrain in the area of operations. What advantages did it give to the attackers or to the defenders?

Compare the opposing forces: In many ways, this is the heart of your study. Describe and analyze the forces involved using the following terms:

> *OAKOC*
> *O*bservation and
>   fields of fire
> *A*venues of Approach
> *K*ey and decisive terrain
> *O*bstacles
> *C*over and concealment
>   *(FM 3-0)*

Size and composition. Which principal combat and supporting units were involved in the operation? What were their numerical strengths in troops and key weapons systems? How did the commanders organize them?

Technology. What were the battlefield technologies—such as tanks, small arms, close air support aircraft—of the opposing forces? Did one side have a technological advantage over the other?

Logistical systems. How did logistics affect the battle? Did one side have an advantage in available supplies or transportation?

Command, control, and communications (C3). What kind of C3 systems did the opposing forces employ? Were these systems under centralized or decentralized control? How were the staffs organized, and how effective were they?

Intelligence. What intelligence was available to the opposing forces? How well was it used? What were the major sources of intelligence? Did one side have an advantage over the other in intelligence resources?

Doctrine and training. What was the tactical doctrine of the opposing forces, and how did they use it? What was the level of training in the opposing forces? Were some troops experienced veterans, some not, and some in-between?

Condition and morale. What was the morale of the troops before the fighting, and did it change after the fighting began? How long had the troops been committed, and how did weather and terrain affect them? Did specific leaders affect morale?

Leadership. Who were the leaders, and how effective had they been in past actions? What was their training and level of experience?

You won't always be able to answer all these questions. But you should go through the list to determine what information you have and where you may need to do more research.

Describe the Action

Describing the battle itself is what most people consider to be real military history. The format below takes a chronological approach to studying a battle. But you shouldn't feel locked into it. If you need to skip a phase in order to examine a specific topic—such as maneuver or logistics—because it is more important to your overall objective, feel free to do so.

1. State the opposing forces' missions: What were their objectives? What missions did the commanders develop to achieve the objectives? Were there other options—such as attacking, defending, or withdrawing—open to the two sides? Were those options feasible?

2. Describe the initial disposition of forces: What were the locations of the opposing forces' units? How were the units deployed tactically?

3. Describe the opening moves of the battle: Examine each side's initial actions. Did one side gain an advantage over the other in the opening phase of the battle?

4. Detail the battle's major phases: Establish a chronology for the battle while examining the actions after the opening moves. Look for key events or decisions that turned the battle toward one side or the other.

5. State the outcome: Who (if anyone) won the battle? Did either side achieve its objectives? Did the battle provide an advantage to the winning side, and what was it? Did the battle have any long-term effects, and, if so, what were they?

Draw Lessons Learned

This is the most important part of battle analysis. In this step, you turn the historical facts of the battle into finished analysis, with lessons to learn and apply today. In trying to distill lessons from the study of any battle, it's important to look at why something happened. To do so, you will look at what caused the outcome. Look for those essential elements that determined the victory or defeat.

The insights, or "constants of war," gained from the study should transcend time, place, and doctrine.

To be a successful Soldier, you must know history.

LTG George S. Patton Jr.

The only right way of learning the science of war is to read and reread the campaigns of the great captains.

Napoleon

The Staff Ride

An excellent method of studying battles and drawing lessons from them is the **staff ride**. Different from battlefield tours, staff rides combine a rigorous course of historical preparation with an examination of the terrain on which an actual battle occurred. The idea behind the staff ride is that you study the battle thoroughly before arriving at the battlefield site—this guarantees thought, analysis, and discussion. A staff ride links a historical event, a systematic preliminary study, and a visit to the actual terrain to produce a three-dimensional battle analysis. You will participate in a staff ride later in the semester; this section gives you the tools to make the most of the experience.

staff ride

systematic preliminary study of a selected campaign, an extensive visit to the actual sites associated with that campaign, and an opportunity to integrate the lessons derived from each

Abbreviated Case Study: The Battle of Kasserine Pass

Many important elements of Army doctrine grew out of lessons learned at the Battle of Kasserine Pass. Those lessons would prove crucial to the Allied invasion of France that began on the Normandy beaches in June 1944. The following section focuses briefly on the unity of command and strategic vision, tactics and maneuver, and leadership of the forces that met there.

T3 tank in the Battle of Kasserine Pass

The US Army that invaded North Africa during World War II, with its British allies in November 1942, was inadequately equipped, under-trained, and inexperienced. Its leaders were not all of the highest quality, and command arrangements proved unequal to the task. Despite heavy losses in Soldiers and equipment, the Allies managed to hold on and turn back a powerful and determined Axis counterattack—those engagements in February 1943 that became known as the Battle of Kasserine Pass.

One historian of that battle has called it a "disaster for the US Army":

About 30,000 Americans engaged in the Kasserine fighting under II Corps, and probably 300 were killed, almost 3,000 wounded, nearly 3,000 missing. It would take 7,000 replacements to bring the units to authorized strengths. The 34th Division under the French XIX Corps at Sbiba sustained approximately 50 men killed, 200 wounded, and 250 missing. II Corps lost 183 tanks, 104 half-tracks, 208 artillery pieces, and 512 trucks and jeeps, plus large amounts of supplies—more than the combined stocks in American depots in Algeria and Morocco (Blumenson, 1986).

Collision in Tunisia

The French authorities in North Africa, after agreeing to a truce, joined the British and Americans who, by then, in accordance with prior plans, had turned eastward from Algeria and entered Tunisia, and were driving toward Bizerte and Tunis, their ultimate objectives. On the way they quickly ran into opposition. Axis troops had entered Tunisia from Italy shortly after [Operation] Torch [the code name of the Allied invasion of North Africa], and eventually a field-army-size force, under General Juergen von Arnim, built up an extended bridgehead covering Bizerte and Tunis in the northeastern corner. Von Arnim sought to prevent the Allies from overrunning Tunisia and also to permit [German GEN Erwin] Rommel's army to finish withdrawing from Libya into southern Tunisia. The Axis would then hold the eastern seaboard of the country. To guarantee their security on the eastern coastal plain, von Arnim and Rommel needed to control the passes in the Eastern Dorsale, a mountain range running generally north and south. Through that chain were four major openings—Pichon and Fondouk in the north and Faid and Rebaou in the south. Von Arnim seized Pichon in mid-December 1942. Toward the end of January 1943, as Rommel settled into the Mareth Line in southern Tunisia, the Axis desire for the other passes initially spurred what developed into the Battle of Kasserine Pass.

M. Blumenson, Kasserine Pass, 30 January–22 February 1943, in C. Heller & W. Stofft, eds., *America's First Battles, 1776–1965*

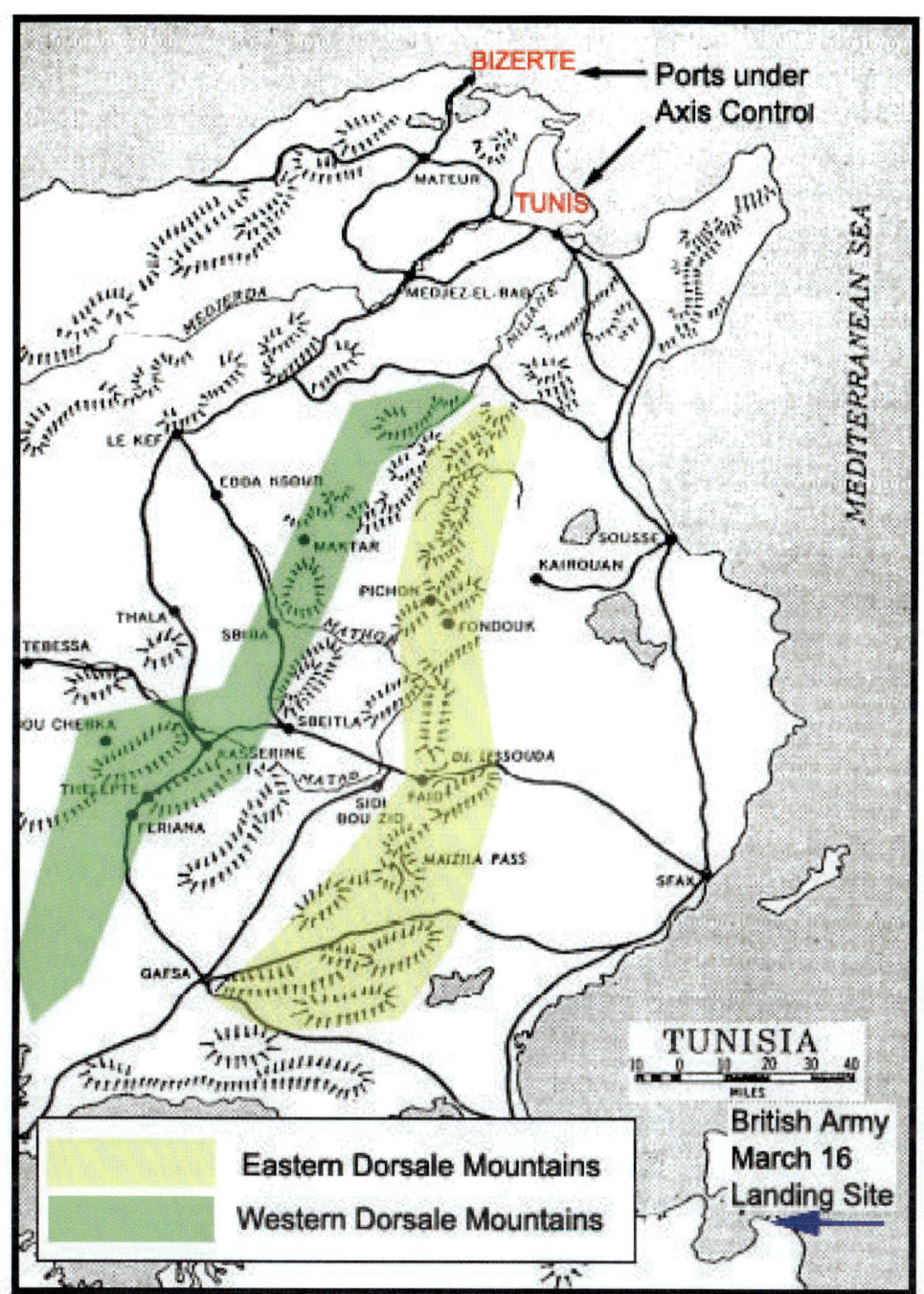

Figure 2.1 Northern Tunisia, Showing the Eastern and Western Dorsale Mountains
Taken from *America's First Battles, 1776–1965*, Blumenson (1986).

Unity of Command and Strategic Vision

In the battle, both sides suffered from divided command arrangements that interfered with the ability of maneuver commanders to carry out operations successfully.

Phase I—Assault on the Eastern Dorsale Begins

Allied command lines were less than firm. General Sir Kenneth A.N. Anderson, at the head of the British First Army—with the British V Corps, several British divisions, and some American and French units in the north—was the overall tactical commander in Tunisia, but Americans found him difficult to work with. [MG Lloyd R.] Fredendall [commander of one of three Allied task forces in North Africa] exacerbated the problem because he saw his role as autonomous. The French, who had General Louis-Marie Koeltz's XIX Corps in the center, a division in the north, another in the south, and miscellaneous detachments scattered virtually everywhere, refused to serve under direct British command. As a consequence, General Alphonse Juin, commander of the French land and air forces in French Northwest Africa, exercised loose direction and provided liaison and guidance to all French formations.

Fredendall had small packets of troops dispersed over a very large area—one battalion of the 1st Infantry Division at Gafsa, another blocking the Fondouk road to Sbeitla, Combat Command A (CCA) of the 1st Armored Division at Sbeitla, Combat Command B (CCB) near Tebessa. He could bolster the French garrisons holding the Faid and Rebaou Passes, keep his forces concentrated in a central location and ready to counterattack, or strike toward the east coast to sever the contact between von Arnim's and Rommel's armies. He sought to do the latter by raiding a small Italian detachment at Sened on 24 January. The action was highly successful as a morale builder but had no real result except to squander Fredendall's meager resources.

The Axis command correctly read the situation and continued planning to take control of the Eastern Dorsale. Rommel established his headquarters in southern Tunisia on 26 January, and two days later *Comando Supremo* in Rome approved a cautious push to take the Fondouk and Faid Passes and to advance on Gafsa. With Rommel's *10th* and *21st Panzer Divisions* temporarily under von Arnim's control, von Arnim attacked on 30 January to open the Battle of Kasserine Pass. Just before dawn, thirty tanks struck 1,000 French troops in the Faid Pass while another contingent of German tanks, infantry, and artillery drove through the Rebaou defile ten miles to the south, overran several hundred French defenders, and came up behind the French holding Faid. Encircled and outnumbered, the French fought gallantly for more than twenty-four hours until they were overwhelmed.

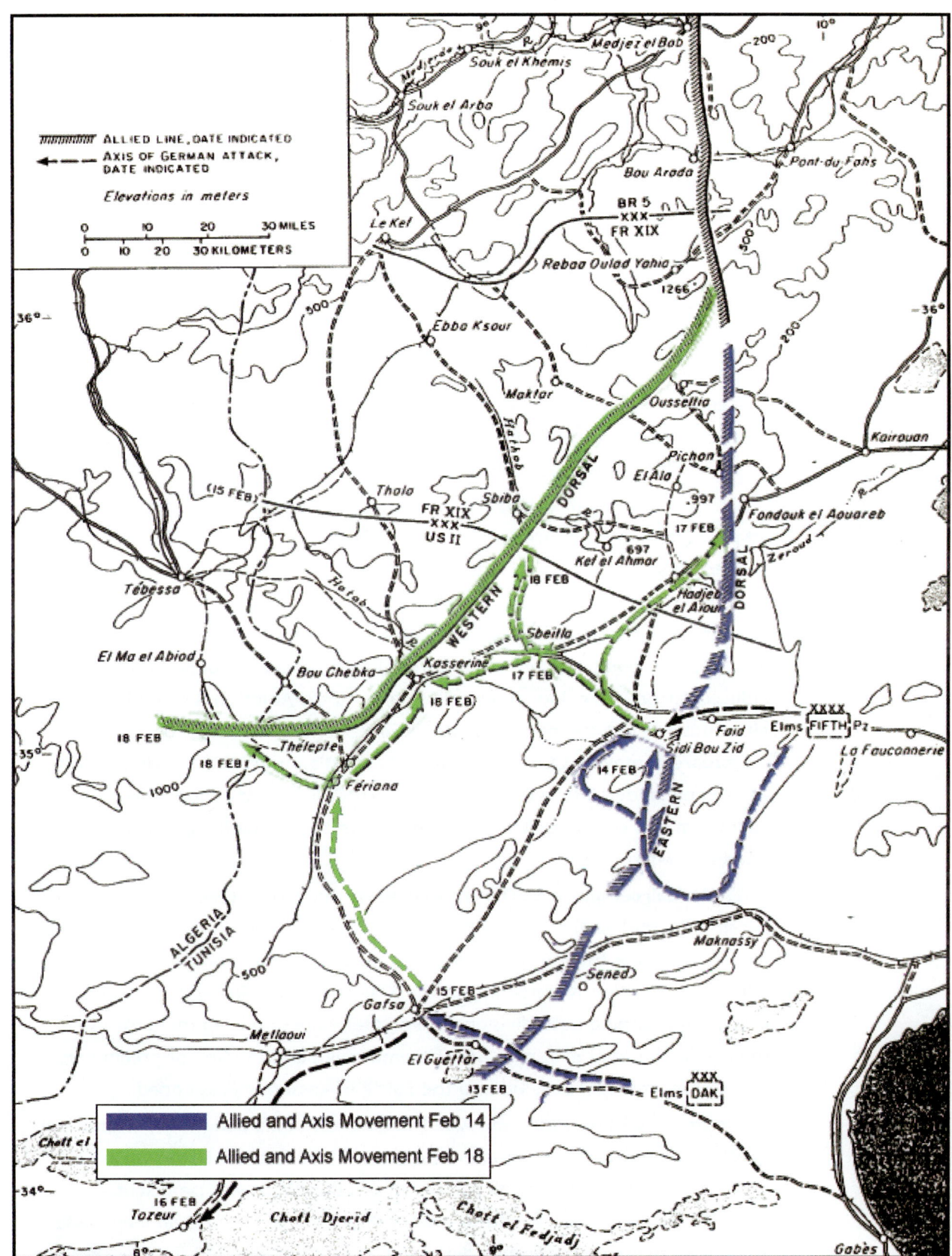

Figure 2.2 Battle of Kasserine Pass, 14–18 February 1943

Five hours after the German attack started, Anderson instructed Fredendall rather vaguely to restore the situation at Faid. Because [American MG Orlando] Ward, the 1st Armored Division commander, was at Gafsa supervising the Sened raid and other useless actions, Fredendall communicated directly with the CCA commander at Sbeitla, Brig. Gen. Raymond McQuillin, who was old in appearance, mild in manner, and cautious in outlook. McQuillin sent out two small reconnaissance units toward the Faid and Rebaou Passes to determine what was happening. At noontime, even though the French at Faid were still resisting, the reconnaissance elements erroneously reported the Germans in control at both passes. McQuillin decided to counterattack. As he moved his assault forces forward, German planes bombed and attacked his units and disrupted the advance. American aircraft dispatched to intercept the Germans dropped bombs on the CCA command post by mistake, and American antiaircraft gunners shot down an American plane. McQuillin then waited for nightfall. During the hours of darkness, he pushed his forces about halfway to Faid and Rebaou.

On the morning of 31 January, more than twenty-four hours after the German attack, McQuillin committed a small-tank infantry force under Col. Alexander N. Stark, Jr., to strike to Faid and another such force under Col. William B. Kern to go for Rebaou. Late getting under way, the effort was badly coordinated and too weak to attain the objectives. Heavy German defensive fires, together with effective bombing and strafing from the air, knocked out several tanks and induced terror, indecision, and paralysis among the American units. McQuillin's effort petered out. As Fredendall, the II Corps commander, was thinking on 1 February of moving CCB from Tebessa to Sbeitla, Anderson, the First British Army commander, instructed him to dispatch CCB toward Fondouk, where von Arnim had struck Koeltz's French elements, seized the pass, and threatened a serious penetration.

Fredendall complied. McQuillin tried again that day to reach Faid but failed because, he said, of the disgraceful performance of Stark's infantry. Von Arnim, now in control of the four major passes, called off further endeavor. With the *10th Panzer Division* at Fondouk and the 21st at Faid and Rebaou, von Arnim, instead of returning both divisions to Rommel, hoped to keep them for use in the north. The front in Tunisia now became quiet, and the first or preliminary phase of what would develop into the Battle of Kasserine Pass ended.

On the Allied side, [Supreme Allied Commander GEN Dwight D.] Eisenhower questioned Fredendall's competence, Anderson doubted the battleworthiness of American troops, Fredendall wondered whether Ward was proficient, McQuillin castigated Stark, and so it went down the line. American ineptitude and failure to rescue the French defenders at Faid had shocked the French.

Additional American units—parts of Maj. Gen. Terry Allen's 1st Infantry Division and of [US MG Charles W.] Ryder's 34th Division—moved into southern Tunisia but they were split into small parcels and physically separated.

M. Blumenson, Kasserine Pass, 30 January–22 February 1943,
in C. Heller & W. Stofft, eds., *America's First Battles, 1776–1965*

The Axis suffered from similar problems, only to a worse degree. The Germans and Italians didn't trust each other. The Axis effort was technically under command of the *Comando Supremo* in Rome, which had to authorize all actions. The two German commanders on the scene, von Arnim and Rommel, nursed petty jealousies and struggled over command of the two panzer (tank) divisions.

Tactics and Maneuver

During the North African invasion, American forces in particular demonstrated serious shortcomings in training, tactics, maneuver, and skill in using their equipment. Some of that equipment, notably American light tanks, was clearly no match for the heavy German Tiger tanks. These problems gave the German-Italian forces—who were battle-hardened and well-trained, and enjoyed close coordination between infantry, tanks, artillery, and close air support—a distinct tactical advantage in the battle.

Phase II—Defeat and Retreat

The second phase of the Kasserine battle started very early on the morning of 14 February, before [COL Thomas D.] Drake could institute his bazooka-firing training program on Djebel Ksaira. [His troops had just received the antitank weapons and did not know how to use them.] During a raging sandstorm, more than 200 German tanks, half-tracks, and guns of both panzer divisions came through Faid. One task force swung around the northern side of Lessouda and encircled the hill; another swung around the southern side of Ksaira and surrounded the height. [LTC John K.] Waters' and Drake's forces, Fredendall's blocking positions, were thus marooned. A series of American mishaps, due largely to inexperience, then permitted the Germans rather easy and quick success. The bad weather relaxed the Americans' security arrangements, and they were unable to react quickly and firmly. Until the storm lifted, men on the hill had difficulty identifying the German elements and held their fire. At 0730, as the weather cleared, McQuillin initiated planned countermeasures. He limply told [US LTC Louis] Hightower to clear up the situation. As Hightower prepared to drive to Djebel Lessouda and relieve the American defenders, enemy aircraft struck Sidi bou Zid and temporarily disrupted McQuillin's command post and Hightower's preparations. Throughout the rest of the day, German planes harassed the Americans. Despite repeated requests for air support, only one flight of four American aircraft appeared briefly over the battlefield.

Hightower went into action with forty-seven tanks. Although outnumbered, he fought bravely against the more effective German tanks. By midafternoon, all but seven of his tanks had been destroyed. During the engagement, some American artillerymen panicked and abandoned their guns. The 1st Armored Division Reconnaissance Battalion, ready to rescue Drake's men on Djebel Ksaira, was unable to even start its counterattack because some of the German tanks surrounding Drake had thrust forward toward Sidi bou Zid and captured a reconnaissance company. The rest of the American reconnaissance units then pulled out and headed for Sbeitla.

With his command post in Sidi bou Zid directly threatened, McQuillin, covered by Hightower's engagement, decided to withdraw to Sbeitla. He phoned and asked Ward to provide a shield by blocking the main road from Faid to Sbeitla. Ward sent Kern and his infantry battalion to take up defensive positions eleven miles east of the town at a road intersection that became known as Kern's Crossroads. Around noon, McQuillin started to move his artillery units and command post out of Sidi bou Zid. German dive bombers attacked them and prompted confusion. As a consequence, for several hours McQuillin lost communications with his subordinate units. That afternoon a swirling mass of American troops—McQuillin's command post, miscellaneous elements, Hightower's remnants, artillery pieces, tank destroyers, engineer trucks, and foot Soldiers—fled toward Sbeitla. McQuillin reestablished his command post there and began to assemble and reorganize his units. Initial estimates of losses on that day were shocking: 52 officers and more than 1,500 men missing. The final numbers of casualties on 14 February were much smaller: 6 killed, 32 wounded, and 134 missing. But between Faid and Kern's Crossroads on the Sbeitla plain, forty-four tanks, fifty-nine half-tracks, twenty-six artillery pieces, and at least two dozen trucks were wrecked, burning, or abandoned. An artillery commander, Charles P. Summerall, Jr., took his men out during the night to recover guns, trucks, and ambulances; on the following morning, he had eight instead of his normal twenty-four pieces—the others were lost—backing the troops at Kern's Crossroads.

In Algiers, Eisenhower ordered American units in Algeria to start for Kasserine Pass, a movement requiring several days' travel. News of their departure, he surmised, would perhaps hearten the troops in Tunisia. While Eisenhower, Anderson, and Fredendall prepared to withdraw to the Western Dorsale, Ward looked forward confidently to his counterattack on 15 February. [COL Robert I.] Stack's infantry and [LTC James D.] Alger's tanks were to marry up at Kern's Crossroads, drive to Sidi bou Zid, then rescue the troops on the heights of Lessouda and Ksaira. While Alger, who had yet to lead his troops in combat, studied the terrain from a hill on the morning of 15 February and Stack readied his infantry for the advance, a flight of German bombers struck their formations and prompted enormous confusion. The counterattack finally started at 1240 in great precision across the Sbeitla plain. Alger's tank battalion led, his three tank companies advancing in parallel columns with a company of tank destroyers, half-tracks mounting 75-mm guns, flaring out on the flanks and protecting two batteries of artillery. Behind rode Stack's infantry in trucks and half-tracks with several antiaircraft weapons as protection. Unfortunately, steep-sided *wadis*— dry stream beds—crossed the plain irregularly and disturbed the careful spacing of the attacking troops. As the tanks crossed the first ditch, German dive bombers jumped them. They bombed and strafed again at the second gully. At the third depression, German artillery began firing. Finally, German tanks emerged from hiding and started to encircle the entire American force. The Americans, fighting bravely and desperately against superior German weapons and experienced German troops, tried to beat back the German wings threatening to surround them. At 1800, Stack ordered all units to disengage and return to Kern's Crossroads. The infantry and artillery escaped relatively unscathed. The tanks were completely destroyed. Alger was taken prisoner, 15 of his officers and 298 enlisted men were missing, and 50 of his tanks had been knocked out. In two days of battle, the 1st Armored Division lost ninety-eight tanks, fifty-seven half-tracks, and twenty-nine artillery pieces.

Just before darkness, a pilot dropped a message from Ward to the troops on Lessouda. They were to get out during the night. Waters having been captured, Maj. Robert R. Moore, who had taken command of the 2d Battalion, 168th Infantry, fewer than two weeks earlier, displayed magnificent leadership and marched out about one-third of the 900 troops on Lessouda to Kern's Crossroads. The other men, together with vehicles and equipment, fell into German hands. Drake on Djebel Ksaira received a message from McQuillin on the afternoon of the following day, 16 February, to fight his way out. That night, Drake led his men off the hill and across the plain. German troops intercepted them and captured almost all. Only a handful reached safety. The two battalions of the 168th Infantry involved on Lessouda and Ksaira sustained losses of about 2,200 men. Two hundred of the Soldiers reported missing were from the southwestern Iowa National Guard units. Meanwhile, when Rommel's attack forces, an Italo-German group of 160 tanks, half-tracks, and guns, learned on the afternoon of 15 February that the Allies had abandoned Gafsa, they advanced to the town, entered, and patrolled toward Feriana. That brought the second phase of the battle to a close.

That evening, *Comando Supremo* gave von Arnim permission to attack Sbeitla, and he jumped off at once. After nightfall, preceded by reconnaissance units, German tanks approached Sbeitla in three columns, firing as they advanced. Shells dropping into Sbeitla prompted McQuillin to shift his CCA headquarters to a location west of the town. Many American troops misinterpreted the movement and believed a wholesale evacuation was in progress. A good part of the CCA defenders panicked and fled. Why?

Night fighting was a new and terrifying experience for most of the men. The solidity of the defensive line was more apparent on a map than on the ground. Because of the darkness, the troops were not well placed. Because of the haste of the withdrawal, they were not well dug in. The harrowing events of three days of defeat had exhausted many Soldiers, morally and physically. Uncertain and nervous, fatigued and confused, hemmed in by widespread firing that seemed to be all around them, believing that the Germans were already in Sbeitla, demoralized by the piecemeal commitment and intermingling of small units, no longer possessing a firm sense of belonging to a strong and self-contained organization, and numbed by a pervading attitude of weariness and bewilderment, many men lost their confidence and self-discipline.

A churning mass of vehicles surged through the town and departed. When engineers demolished an ammunition dump, they intensified fear and prompted additional departures. Around midnight, concerned over his ability to hold Sbeitla, Ward telephoned Fredendall and suggested reinforcing Kasserine in strength.

M. Blumenson, Kasserine Pass, 30 January–22 February 1943, in C. Heller & W. Stofft, eds., *America's First Battles, 1776–1965*

Leadership

As the battle progressed, it became clear to the Allies that stronger command was required. The British general in charge of Allied forces in Tunisia and deputy Allied commander, British MG Harold Alexander, conferred with Eisenhower and visited the Allied sectors. On the Axis side, Rommel, a leader of strong will and highly respected by his troops, moved to the front to command his forces directly. But in the end, his indecisiveness in choosing a point of attack, his own fatigue and that of his soldiers (Rommel still had to worry about British GEN Bernard Montgomery's Eighth Army approaching from the east), and an inability to coordinate with von Arnim led the German-Italian troops to break off the attack and withdraw.

Phase III—The Line Holds

On the Allied side on 18 February, the shock of defeat was visible among the troops. Everyone was tired. Units were mauled, dispersed, and mixed; had no specific missions; lacked knowledge of adjacent formations. The troops seemed to be slipping out of control. Eisenhower sent artillery and tank destroyers from Algeria to Tunisia. A shipment of 295 new Sherman tanks had just arrived, but unwilling to risk losing them all, he released 30 to the British and 30 to the 1st Armored Division. Alexander had come to Algiers on 15 February in accordance with agreements reached at the Allied Casablanca Conference in January and prepared to take command of the ground forces in Tunisia— Anderson's First Army and Montgomery's Eighth—which were approaching the Mareth Line. Alexander conferred with Eisenhower, then toured the British front on 16 February, visited the French sector on 17 February, and traveled on 18 February to the II Corps area. He was horrified to see the state of confusion and uncertainty and was upset by the absence of a coordinated plan of defense. Instead of waiting to take command of the ground forces on 20 February, he assumed command on the nineteenth and ordered everyone to hold in place. There was to be no withdrawal from the Western Dorsale.

[US COL] Anderson Moore's 19th Engineers had been laying mines between the village of Kasserine and the pass, five miles beyond. On 18 February, having covered the withdrawal of CCB through the village and the pass, Moore moved his men through the pass and organized defensive positions. Just beyond the pass, on the western side, the road splits: one route leads to the west toward Tebessa; the other, the main road, goes north to Thala. Moore, with about 200 engineers and infantrymen armed with small arms and automatic weapons and supported by two batteries of US 105-mm howitzers, a battery of French 75s, and a battalion of tank destroyers in the rear, covered the road to Tebessa. An infantry battalion defended the road to Thala. Most of the troops were inexperienced and nervous. On the evening of 18 February, Anderson instructed Koeltz to dispatch a brigade of [British MG Sir Charles] Keightley's 6th Armored Division from Sbiba to Thala. [French] Brig. Charles A.L. Dunphie's 26th Armored Brigade moved. He was thus in place to help the American battalion defending the road from Kasserine to Thala. Or he could move back to Sbiba if the main German threat developed there.

The Casablanca Conference

US President Franklin D. Roosevelt and British Prime Minister Winston Churchill met in Casablanca, Morocco, in January 1943 in their fifth war conference. At the meeting, the two Allied leaders agreed to demand the unconditional surrender of the Axis powers (Germany, Italy, and Japan). They also appointed American GEN Dwight D. Eisenhower as Supreme Allied Commander and named British MG Harold Alexander as his deputy and commander of Allied forces in Tunisia. They forced the Free French leader De Gaulle and the French North African commander to negotiate (unsuccessfully) and made important decisions about aid to the Soviet Union and China, which were also fighting off Axis invasions.

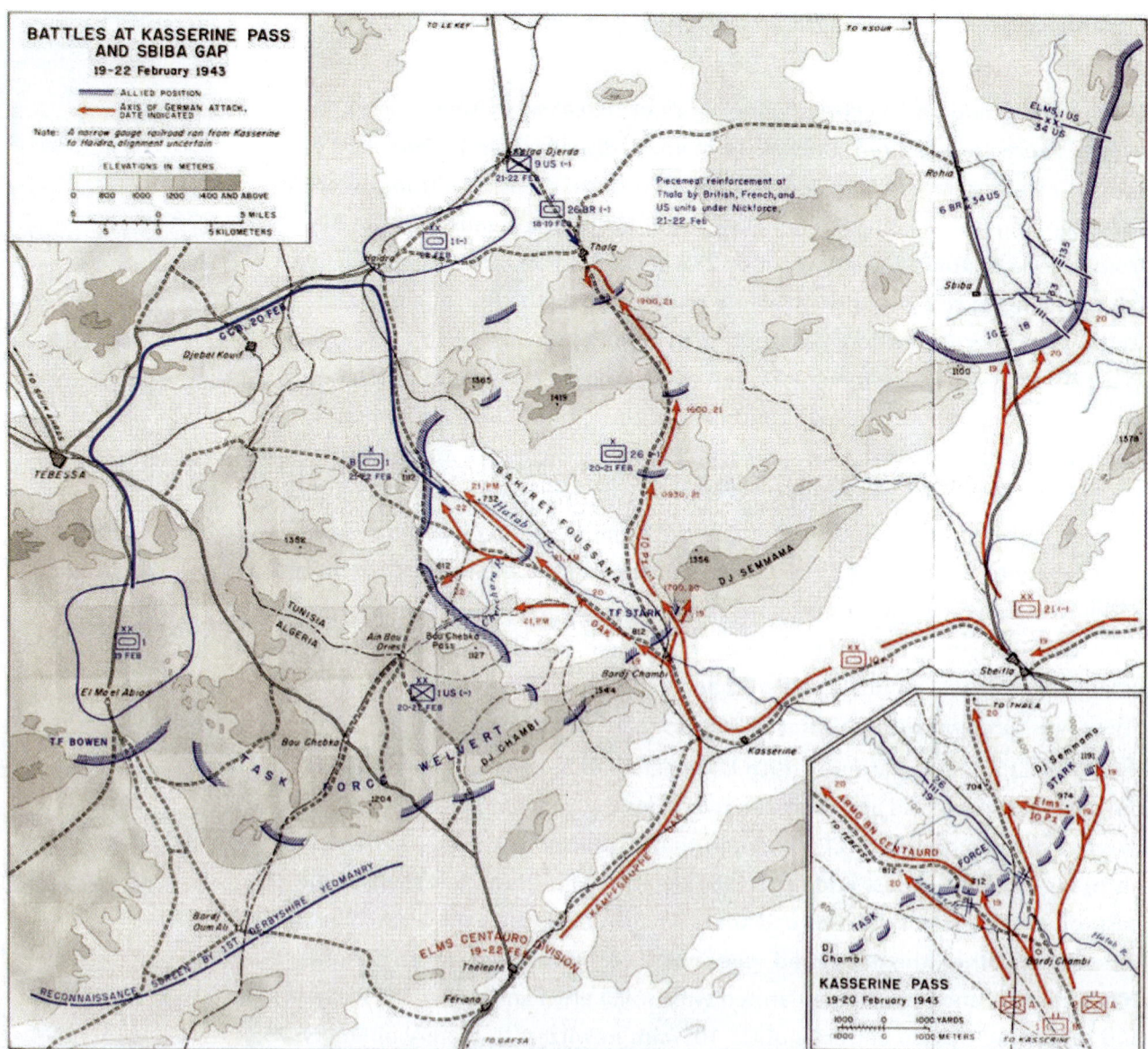

Figure 2.3 The Battle of Kasserine Pass, 19–22 February 1943

Meanwhile, CCA of the 1st Armored Division, having given Keightley's and Ryder's forces, as well as the French, time to set up defensive positions at the Sbiba Pass, drove through Sbiba to Tebessa. On 19 February, CCA arrived at the three minor passes south and west of Kasserine to bolster remnants of a French division, two American battalions (one of Rangers, the other of infantry), the Derbyshire Yeomanry, and CCB. When German reconnaissance units probed the Kasserine Pass on the evening of 18 February, some of Moore's engineers fled. That night Fredendall put Stark in command of all the units defending the pass. Stark arrived on the morning of 19 February as the Germans attacked in earnest.

Seeking surprise, an infantry battalion of the *Afrika Korps* [Germany's veteran forces in North Africa, under Rommel's command] advanced through Kasserine Pass without artillery preparation. When the troops met opposition, a panzer grenadier battalion backed by an 88-mm cannon reinforced them. A unit of British mortars and some reconnaissance elements had just arrived at the Kasserine Pass, and they helped the Americans hold off the Germans. When Moore asked for more infantry to support his engineers, Stark seized on a battalion of the 9th US Infantry Division that had just arrived from Algeria. Stark sent two rifle companies to Moore—one for each flank of Moore's defenses—and kept one for the Thala road, thereby splitting the battalion.

Rommel himself came to Kasserine, was impressed by the opposition, and decided to take his main effort toward Sbiba. But he wished the attack at Kasserine to continue. After clearing the pass, his troops were to strike westward toward Tebessa in order to stretch the Allied defenses. The *21st Panzer Division* had attacked Sbiba that morning, but Koeltz, Keightley, and Ryder had stopped the thrust. Rommel then changed his mind and decided to concentrate in the Kasserine area. He ordered the *10th Panzer Division,* which was on its way to Sbeitla, to continue on to the Kasserine Pass. The division was at half strength, for von Arnim had refused to release some units, particularly the heavy panzer battalion, which had about two dozen enormous Tiger tanks. Because the *10th* was moving slowly, an impatient Rommel brought up the [Italian] *Centauro Division.* He now wished the *Afrika Korps* to open the pass and to drive westward toward Tebessa. The *10th Panzer Division,* after going through the Kasserine Pass, was to strike at Thala. That evening, the 16th Infantry of the 1st Division marched from the Sbiba area to the Kasserine area. Fredendall sent it to bolster the minor passes south and west of Kasserine. He gave General Allen, the 1st Division commander who was with the regiment, the job of coordinating the defenses of these passes. Fredendall then ordered CCB of the 1st Armored Division to back up the engineers on the Tebessa road at Kasserine Pass where the defenses seemed on the verge of collapse. Dunphie, commander of the 26th Armored Brigade at Thala, asked permission to reinforce Stark, but Keightley wanted him to be on hand if he was needed at Sbiba. Dunphie nonetheless sent eleven of his tanks from Thala to buttress Stark's positions that night.

On 20 February, the *21st Panzer Division* attacked Sbiba again and made no progress. But at Kasserine, the shrieks of the *nebelwerfer,* multiple rocket launchers that had been recently introduced by the Germans, unnerved Moore's engineers holding the Tebessa road. They fell apart, and by afternoon—having lost eleven men killed, twenty-eight wounded, and eighty-nine missing in three days (and many more had temporarily vanished)—they no longer existed as a coherent force. Fortunately, [BG Paul] Robinett's CCB arrived and blocked the road.

On the main route to Thala, although jittery, the defenders held. Rommel then became even more impatient for a quick victory at Sbiba and Kasserine. He was apprehensive over the Mareth Line positions, for Montgomery had just that day attacked his outposts in southern Tunisia. Late in the afternoon, on Anderson's order, Keightley dispatched Brig. Cameron Nicholson, his assistant division commander, from Sbiba to Thala with miscellaneous troops. No longer confident of Fredendall's ability, Anderson wished Nicholson to command, as Fredendall's representative, all the British, American, and French fighting on the west side of Kasserine Pass. What actually developed was that Fredendall and Robinett commanded the forces blocking the Tebessa road, and Nicholson and Dunphie took control of the units defending the Thala road.

On 21 February, Rommel let the attacks in the Sbiba area continue but looked for decisive success at Kasserine. He decided to make his main effort to Thala and to head for Le Kef beyond. Furious fighting on both the Tebessa and Thala roads resulted in a slight German advance toward Tebessa and the prospect of German tactical success at Thala. By now, Stark's force on the Thala road had virtually evaporated, and Dunphie emerged as the chief Allied protagonist. Committing his tanks and infantry against a strong thrust directed by Rommel himself, who took control of the battle for several hours, Dunphie lost the bulk of his armor and had to withdraw to the final line of defense before Thala. The Germans followed, and fierce combat erupted after darkness and ended in a draw. Both sides retired 1,000 yards—Dunphie to the north, the Germans to the south. The final defensive line was virtually uncovered, and Rommel seemed about to enter Thala. Expecting just that, Anderson asked Koeltz, who had again stopped the Germans at Sbiba, to send a battalion of infantry and whatever else he could to Thala. Because Ryder was making some local adjustments, Koeltz requested Keightley to dispatch elements. That night, a battalion of British infantry and some tanks traveled along a mountain trail to reinforce Nicholson and Dunphie.

Meanwhile, Allied units were coming from Algeria. A battalion of French infantry moved from Constantine and arrived at Sbiba. Fifty-two Sherman tanks and crews were en route to Tebessa. A provisional British unit with twenty-five new Churchill tanks reached Sbiba. The 47th Infantry of the 9th US Division was on the way from Oran to Tebessa. Most important, Brig. Gen. S. LeRoy Irwin's 9th Division Artillery, with three artillery battalions and two cannon companies, traveling from western Algeria, got to Tebessa on the afternoon of 21 February. Ordered to Thala at once, Irwin's guns were in position by midnight. Nicholson placed Irwin in charge of all the artillery at Thala, and Irwin sited his forty-eight pieces, plus thirty-six other guns of various calibers, to cover the all-but-abandoned final line of defense, manned now by British infantry reinforced by stragglers rounded up by Stark, about twenty tanks of Dunphie's brigade, plus the British infantry battalion and a few tanks, some of them new Shermans released by Eisenhower, coming from Sbiba. Less than a mile away were at least 50 German tanks, 2,500 infantry, 30 artillery pieces, and other weapons, including the notorious *nebelwerfer.*

The *10th Panzer Division* was ready to start what Rommel expected would be the advance into Thala on the morning of 22 February, when Irwin's guns opened up. Expecting a counterattack, the Germans postponed their effort. Nicholson launched a foray and, although he lost five tanks, bluffed the Germans. Rommel came up the Thala road, noted the increased volume of Allied shelling, and gave permission to delay the offensive. Now Robinett and his CCB seemed about to be overwhelmed. During the previous night, approximately a battalion of German and Italian troops had infiltrated the American positions. Intending to strike toward Tebessa, they became lost. On the morning of 22 February, they arrived in the rear of the miscellaneous Allied troops—American, French, and British— guarding the Bou Chebka Pass, one of the minor defiles south and west of Tebessa. The Axis force captured several American howitzers and antiaircraft guns and prompted considerable anxiety over the security of that pass and two others nearby. It took most of the day to track down, disperse, and capture the Italo-German unit. Under the impression that Allied defenses were caving in, Fredendall went to the commander of the under-strength French division in the area and asked him to defend Tebessa. While Fredendall was gone, someone at the II Corps headquarters decided to move the corps command post to avoid being overrun. When Fredendall returned, he found his headquarters half abandoned; many clerks and radio operators were on the way to Le Kef and Constantine. Feeling unable to maintain control, Fredendall, having already passed responsibility to Allen for the minor passes, now instructed Ward to coordinate the defenses on the Tebessa road. Learning that the 47th Infantry of the 9th Division was about thirty miles south of Constantine, Fredendall asked the regiment to remain where it was in order to protect Constantine in case the Axis forces broke through Thala and Tebessa.

During the night of 22 February, Anderson, whose British First Army headquarters was nine miles north of Sbiba, shifted his command post behind Le Kef. Koeltz almost pulled his headquarters back too, for von Arnim had attacked half-heartedly in the Pichon area. But Koeltz drew Keightley's and Ryder's divisions out of Sbiba and faced them toward Thala to meet the expected breakthrough there. Sbiba lay open to German entry. However, nothing happened at Sbiba or at Kasserine. After conferring with [German Field Marshal Albert] Kesselring, who came to Tunisia on the afternoon of 22 February, Rommel called off his attack. [Kesselring, stationed in Rome, was the de facto German theater commander for the Mediterranean.] He [Rommel] had been unable to secure von Arnim's cooperation. He thought it impossible to obtain a decisive victory before Montgomery attacked the Mareth Line. His units were fatigued, and Rommel himself was extremely tired and discouraged. That night, Rommel ordered his forces to withdraw to the Eastern Dorsale and the east coast. They did so early on the morning of 23 February, leaving a profusion of mines and destroyed bridges in their wake. There was no Allied pursuit of the departing enemy. According to Koeltz, the Allied units "were in such disorder and their commanders so shaken" that no immediate reaction was possible. The Battle of Kasserine Pass was over.

M. Blumenson, Kasserine Pass, 30 January–22 February 1943, in C. Heller & W. Stofft, eds., *America's First Battles, 1776–1965*

Lessons Learned

The significant losses in men and materiel during the Battle of Kasserine Pass led to a harsh reappraisal by US commanders of American training, tactics, and equipment, which resulted in significant changes in Army doctrine and field manuals. But the battle also exposed serious Axis weakness as well.

In the areas of *unity of command* and *strategic vision*, the lack of trust between the Germans and the Italians, who were unable to work as effectively in coalition as the British, Americans, and French, impeded operations. Technically fighting under Italian command, Rommel and von Arnim had to get *Comando Supremo* approval for their overall plans. The disputes and rivalries between the two German generals only made matters worse. "Had Axis forces been closely coordinated by an overall commander in pursuit of bold objectives enunciated by a self-confident coalition, the Axis would, no doubt, have attained a strategic victory instead of merely a tactical success" (Blumenson, 1986).

The Allies had their own disputes, of course. The French still deeply resented the British destruction of their fleet in Algeria after the fall of France in 1940 and quarreled among themselves. For their part, the British were not impressed by the American performance on the battlefield, or by the leadership of generals such as Fredendall. Command lines between and among the Allied forces were fuzzy at best. Nevertheless, the Allies were able to work together well enough to blunt a determined Axis offensive; the command team of Eisenhower and Alexander (who would later become a field marshal) proved particularly effective in appraising the situation and providing the necessary support.

In *tactics* and *maneuver*, the veteran German and Italian forces' experience, coupled with their superior equipment (especially in tanks and rockets) clearly gave them the edge in the battle. They also demonstrated the effectiveness of tactical air support of ground operations, and they benefited from a numerical advantage of two armies to one on the battlefield (before Montgomery's Eighth Army arrived from the east).

Estimated Losses at Kasserine Pass

Killed
Germans—200
Americans—350

Wounded
Germans—550
Americans—3,200

Missing
Germans—250
Americans—3,000

Equipment
Germans—
14 guns,
61 motor vehicles,
6 halftracks, and
20 tanks
Americans (II Corps)—
183 tanks,
104 halftracks,
208 artillery pieces,
512 trucks and Jeeps

(Italian losses for the battle are unknown.)

The Americans, on the other hand, demonstrated inexperience with their equipment, some of which—many tanks, for example—was seriously inferior to the Germans'. US Soldiers often could not distinguish enemy from friendly equipment on the battlefield. The Americans received little direct support from their armor or friendly aircraft (which also caused numerous Allied casualties). Too many US planes bombed their own forces—if their own forces didn't shoot them down first. Not until after the Normandy invasion the following year would Allied pilots be able to talk directly to the ground units they were supporting or would tactical air commands work closely with field armies. Likewise, infantry and armored units would not train together until late in 1943.

Allied commanders lacked the ability to coordinate units in battle, set unit boundaries, mass their fires, and handle traffic on the roads or prisoners of war. They dispersed units and tanks and employed them piecemeal rather than concentrating their firepower. The defensive attitude American Soldiers adopted undermined "the offensive spirit by which alone we can win battles," in one general's informed opinion. Allied forces were so shaken by the battle that it took them two days to figure out that Rommel had withdrawn from the battlefield. America had been unprepared for war when war came, and Americans at Kasserine "paid in blood the price of battlefield experience."

The battle likewise taught grim lessons regarding *leadership*. The Allied commanders' orders were vague and imprecise. Fredendall was arrogant, opinionated, and "perhaps less than stable," while McQuillin and Stark's reactions were "slow, cautious, and characteristic of World War I operations." On several occasions, senior US generals lost contact with subordinate units or their own headquarters. Commanders at the front had to relay too many questions to higher headquarters, with decisions often coming back after it was too late. Despite these shortcomings, however, several individuals—Ward, Robinett, Hightower, and others—demonstrated solid leadership ability.

The two sides studied the results of this first contact between US and German forces carefully. Rommel developed a low opinion of US tactics and equipment—an appraisal that would serve him poorly a year later in France. For in the face of a tactical defeat, the American forces recovered and showed they could learn from their mistakes. After Kasserine, their competence and confidence grew, they received better equipment and used it more effectively, and they developed a better ability to work with their allies (Blumenson, 1986).

The Americans made many mistakes in this first large-scale engagement of the war in Europe, but they learned from their errors and made adjustments that enabled them to go on to victory in Tunisia and beyond. The defeat at Kasserine showed the Army what troops had to learn and to do. That they quickly became proficient in the warfare of the 1940s confirmed their spirit, their flexibility, their strong sense of purpose—their will to win (Blumenson, 1986).

Critical Thinking

Which factor do you think played the biggest role in the Axis' initial victory at Kasserine Pass—unity of command and strategic vision, tactics and maneuver, or leadership? Which factors later allowed Allied powers there to check the German advance? With all the challenges they faced, why did the Allies succeed in later turning defeat into victory in North Africa?

CONCLUSION

The purpose of battle analysis and the study of military history—in war and in peace—is to learn the important lessons the past has to teach today's Soldiers. The study of military history also allows you to understand military concepts, study the lives of Soldiers in the past, understand how doctrine has evolved, broaden your military knowledge, understand leadership issues, and learn about strategy and tactics.

The study of military history should be an integral part of your career as an Army officer, regardless of rank or years of service. You'll find that as you grow in experience, your study of military history and your skill at battle analysis will become more and more relevant to the challenges you and the Army face.

Key Words

battle analysis
staff ride

Learning Assessment

1. Describe the four steps of battle analysis.
2. Explain the purpose of battle analysis.

References

Anderson, C. R. (n.d.). *The US Army Campaigns of World War II: Tunisia.* Washington, DC: US Army Center for Military History. Retrieved 6 December 2005 from http://www.army.mil/cmh/brochures/tunisia/tunisia.htm

Austerman, W., & Poyas, F. (n.d.). *The Battle Analysis Paper and Presentation: AMEDD Officer Advanced Course.* Retrieved 25 November 2008 from http://www.cs.amedd.army.mil/History/ba/intro.html

Blumenson, M. (1986). Kasserine Pass, 30 January–22 February 1943. In C. Heller & W. Stofft (Eds.), *America's First Battles, 1776–1965* (pp. 226–265). Lawrence, KS: University of Kansas Press. Retrieved 5 January 2005 from http://www.army.mil/cmh/books/Staff-Rides/kasserine/Vol-I-Part_1.pdf

Blumenson, M. (1987). Constants of War: The Relevance of Kasserine. *Army, 37.*

Field Manual 3-0, *Operations.* 27 February 2008.

Field Manual 5-0, *Army Planning and Orders Production.* 20 January 2005.

Robertson, W. G. (1987). *The Staff Ride.* Washington, DC: US Army Center for Military History. Retrieved 6 December 2005 from http://www-cgsc.army.mil/carl/resources/csi/robertson/robertson.asp

TSP 155-H-0010, *Integrate Historical Awareness and Critical Thinking Skills Derived from Military History Methodologies into the Training and Education of Self and Subordinate Leaders.* 1 January 1999.

CULTURAL AWARENESS

Key Points

1 Basics of Organizational Culture

2 Levels of Culture

3 Sources of American Culture

4 Obstacles to Cultural Understanding

5 Culture and Military Operations

The most important way to avoid social blunders
is to show respect for the dignity of the individual
and his or her way of life.

University of Military Intelligence, Iraq: Culture Overview

Introduction

Culture is a shared system of knowledge, behavior, and beliefs that unites members of a society or group. Culture is learned. People, organizations, and countries pass down their culture to successive generations to ensure its survival.

Soldiers come into the Army with knowledge, behaviors, and belief systems that are unique to the particular group of which they are a part. To be an effective leader, you must understand and respect the cultural backgrounds of your Soldiers. If you respect them, they will respect you. They will also be more responsive to your commands.

As an Army lieutenant, you will travel to many places in the world and interact with people of many cultures, both in war and in peace. For this reason, it is also essential that you understand the culture of the country you are operating in and that you help your Soldiers achieve a similar understanding. If you and your Soldiers aren't aware of and sensitive to the people's culture and customs, you may easily insult, anger, or alienate them. You may do this without even knowing it.

A failure to understand the culture of your host nation or enemy may put your unit's mission in jeopardy. However, if you and your Soldiers understand what people of the country you are in value—what they believe and how they interact and communicate—you will be able to predict and respond to that group's behavior more effectively. You will be more likely to accomplish your mission.

culture

a shared system of knowledge, behavior, and beliefs that ties together a society or group and gives a perspective on the world and a specific outlook on life

Basics of Organizational Culture

In this section, you will learn about the concept, influence, and levels of organizational culture. As a leader, you need to understand these concepts so that you can apply them both to your own organization and to the Contemporary Operating Environment (COE). Understanding the people living in your area of operations (AO) and their culture(s) is important to your counterinsurgency operations (COIN). It's also important to remember that cultural effects are not limited to one unit's AO. At a tactical level, you must be aware of activities in neighboring regions and population centers that may affect the population in your AO.

Make no mistake—if your Soldiers do not understand, and are thereby more prone to show disrespect for, another culture, they may seriously impede your unit's mission. Military personnel with a superficial or inaccurate picture of a host culture can make enemies for the United States. Officers and Soldiers who understand the host culture, by contrast, can earn the local people's respect. Likewise, to function as a team and achieve their objectives, your Soldiers must appreciate one another's unique cultural backgrounds.

To advance your unit's mission and your own success as an officer, you must be aware of culture—the "why behind the what"—in your unit and in the countries of the COE.

How easy is giving offense in another country?

A gesture that is friendly or innocent in one culture may have a very different meaning in another. For example, in the Iraqi culture showing someone the soles of your feet—or the bottoms of your shoes—is a serious insult. American hand signals such as the "OK" or "thumbs up" gesture are considered obscene in Iraq.

All too often, cultural misunderstandings or a sense of cultural superiority has led to military disaster, as the British army learned during the Zulu wars in nineteenth-century South Africa.

Military Disaster in South Africa

Lord Chelmsford, the British commander in South Africa in 1879, held the native Zulu in contempt. Overconfident and completely ignorant of the Zulu warrior ethos, he aggressively sought battle on poor terrain and completely disregarded the advice of the Boers, who had considerable experience fighting the Zulu. The ensuing Battle of Isandlwana would be the worst defeat ever inflicted upon the British at the hands of an indigenous people. Despite having a considerable advantage in firepower, more than 1,300 of Chelmsford's 1,800-man command were killed by 20,000 Zulu mostly armed with spears and clubs. Hearing of the defeat, Lord Chelmsford said, "I can't understand it, I left a thousand men there."

In contrast, just two days later, a smaller detachment of Chelmsford's command, the 2nd Battalion, 24th Warwickshire Regiment, garrisoned at Rorke's Drift (a farm turned into a mission) successfully repulsed an attack of 4,000 Zulu with a total of 135 men. The officer in charge heeded the advice of not engaging the Zulu in the field and instead fortified his position. In all, 15 British Soldiers were killed, while Zulu casualties numbered more than 600 warriors. Eleven of the defenders won the Victoria Cross—Britain's highest military decoration.

Lord Chelmsford's mission failed because he lacked an understanding of Zulu warrior culture. He did not know that his adversary was willing to absorb a tremendous amount of punishment. It's a sound tenet of warfare to thoroughly understand your adversary. Understanding your adversary's culture will enable you to better predict his courses of action—and cripple his power base.

The American military man is a perfect target. He is a symbol of America's military interests overseas. His death weakens the ties between America and our country's military rulers. The American military people are paid to risk their lives for their country. Do not hesitate to kill them. Kill their wives and children if necessary. Make America order them back home or risk open rebellion in the streets of Washington. American military members are highly visible targets. They seem to intentionally act in such a way to be culturally obnoxious and alienate themselves for no apparent reason.

Hussein Balkir, Turkish Workers and Peasants Liberation Army

American Soldiers are highly visible targets overseas.

Levels of Culture

artifacts

things that one sees, hears, and feels when encountering a group culture

espoused values

beliefs and values held by one or more members of a cultural group

shared basic assumptions

attitudes, beliefs, values, methods, or behaviors that have repeatedly enabled the cultural group to solve important problems and that members of the group accept as reality

Organizational leadership expert Dr. Edgar Schein has identified three levels of culture: **artifacts, espoused values**, and **shared basic assumptions**. These levels can be applied to the culture of virtually any organization or country. Once you become familiar with these three levels, try to relate them to your interactions with Soldiers in your unit whose backgrounds are different from your own, as well as to interactions with people from other countries.

Artifacts. Artifacts may be tangible or intangible. They include behaviors as well as physical objects. Examples of artifacts include:

- Architecture: office layouts, decorations, displays, color schemes
- Language, manner of address, idiomatic speech
- Interpersonal behaviors, rituals, ceremonies
- Myths and stories relating to the culture.

Espoused values. Espoused values are beliefs and values that have been adopted by some members of a group—especially its leaders—but that are not necessarily or completely shared by members of the group as a whole.

Shared basic assumptions. Shared basic assumptions are attitudes, beliefs, values, methods, or behaviors that have stood the test of time for a group. Because of this repeated success, members come to believe that the assumptions reflect reality. For these people, the assumptions that they share represent the correct and natural way to think or behave. If a behavior or a belief contradicts a shared basic assumption, members of the culture automatically believe that it is inappropriate or wrong.

Of the three levels of culture Dr. Schein described, the shared basic assumption level has the greatest influence on organizational behavior. It's also the level most resistant to change. Organizational culture solidifies a group's resistance to change. The group unites around its shared basic assumptions and gradually develops a unique identity as a result of members' shared history and the human need for stability, consistency, and meaning.

Shared basic assumptions have great power. For example, you've probably heard someone justify a certain approach to a task by saying, "We've always done it this way." As a group faces challenges, it tries various solutions. When a solution works, the group tends to use it again when faced with a similar situation. If such a solution repeatedly works, it becomes embedded as part of the culture—in other words, it becomes a shared basic assumption. Group members no longer wonder whether the solution will work or think of alternate solutions; they assume that the solution represents a truth of nature.

In other words, the group accepts these shared basic assumptions unconsciously. It is this aspect that anthropologist Shirley Teper referred to when she defined culture as "a habit system in which 'truths' that have been perpetuated by a group over centuries have permeated the unconscious."

Critical Thinking

What are some shared basic assumptions in your family? In your hometown? Among your classmates?

Critical Thinking

What specific behavior have you recognized in another Cadet that could become a modeling behavior in creating a unit culture?

Traumatic or critical events can also create and change culture. In such cases, the successful behaviors of the leader and other key group members may be taken by other members as the correct way to respond to similar crises or situations. These behavioral examples may then become part of the organizational culture. They become shared basic assumptions. If you were to study every president of the United States, you could argue that each one has contributed his own stone—however small—to the overall cultural foundation upon which the United States rests.

Cultural Sensitivity

Cultural sensitivity is awareness of, and respect for, the norms of a culture—particularly one that is different from your own.

Showing respect for a different culture does not mean that you share its values. You do not even have to like it. It simply means that you understand that behavior has cultural roots. Once you become aware that people behave the way they do for reasons that are tied to their history and traditions, rather than individual idiosyncrasies, it becomes easier to understand them and to interact with them in a productive way.

Becoming culturally sensitive is not easy. This is because culture is in many ways like an iceberg. Just as an iceberg has a visible section above the waterline and a larger, invisible section below it, culture has both observable and nonobservable aspects. The visible part of culture—for example, a person's behavior—is only a small part of a much bigger whole. Beneath-the-surface values and assumptions influence the surface behavior.

Other Characteristics of Culture

In learning cultural sensitivity, it helps to keep these characteristics of culture in mind:

Culture is learned. Although culture is passed from generation to generation, it is not something you are born with. It is a learned behavior. The primary means of this transfer is language, and a shared language can be a key unifier for people of a particular culture.

Cultures share many basic traits but express them in different ways. While each culture is unique, all cultures share certain traits. Eating breakfast, for example, is a tradition in all cultures. The foods that people consider appropriate for breakfast, however, differ considerably. All cultures produce art, but the varieties of artwork they produce are incredibly varied. Thus, culture determines not only what we eat for breakfast, but also how we dress, cross our legs, or greet a visitor, as well as what music we like to listen to and our tastes in art and design.

Critical Thinking

How have cell phones changed American culture? How did the attacks of 9/11 change American culture?

Culture is a group survival mechanism. Humans are social animals. They gather together—in families, clans, tribes, regions, states, and even multinational corporations—for companionship and survival. The environment in which people live influences culture by dictating the survival strategies the group needs to flourish. The impulse to survive and thrive leads people to develop distinctive techniques and tools in a variety of social environments and climates.

Cultures are constantly evolving. All cultures change; some do so more rapidly than others. Some of what is taught can be lost. New technology and ideas, as well as significant historical events, can cause rapid cultural change.

Sources of American Culture

While it's useful to know what Americans believe in and value, it's equally important to know why Americans believe what they do—in other words, to understand that American values and beliefs are a result of the national experience. Understanding American culture requires an understanding of US history. Once you understand the culture and history of the United States, you can better appreciate the importance of culture and history to people of other countries.

The essence of cross-cultural understanding is knowing how your own culture is both similar to and different from the local or "target" culture. For this reason, before you pursue cross-cultural knowledge, you must first look at yourself.

People from other cultures are "different" not by nature, but simply in relation to a particular standard against which you're measuring them. To see the differences, therefore, you have to examine that standard. In the case of the military, that standard is the American culture that Soldiers and other service members come from.

You might wonder why people from the United States would need to have their culture revealed to them. Ironically, the key characteristics of American culture may be more evident to people from other groups than to Americans themselves. This is true of any national group. As noted above, culture is deeply ingrained and operates on a subconscious level. As a result, the members of a culture are, in many ways, the least able to see it. Although they embody the culture, they would in fact have to get out of that body if they wanted to see what it looked like. In that sense, you might want to think of this as an "out-of-body experience."

No one American is quite like any other American, but a handful of core values and beliefs underlie and permeate the national culture. These values and beliefs don't apply across the board in every situation; Americans may, on occasion, even act in ways that directly contradict or flaunt them. Nonetheless, they are at the heart of the American cultural ethos.

Some of the major sources of American culture are these:

- *Calvinist Protestantism*—a strong work ethic—the idea that work is intrinsically good—and the notion that salvation is apparent through worldly success; also the idea that no authority figure stands between the individual and God

- *Geography*—the concept of the frontier, unlimited resources and opportunity, isolation, sparse population, distance from Europe and Asia

- *Freedom and independence*—from religious, political, and economic repression and from a rigid class system and social stratification; a strong attachment to representative government

- *The melting pot*—*E pluribus unum*—out of many people or states, one nation; a mixture of many cultures, religions, ethnic groups, and races united by a common commitment to freedom; immigrants who were dissatisfied with their lot in life and willing to take risks to better themselves; a sense of adventure.

Critical Thinking

Based on your present knowledge, how do you think core American values might differ from those of another national culture in the COE?

Critical Thinking

How does the culture of Salt Lake City differ from that of New York City? Why? How do you explain the cultural differences between Charleston, South Carolina, and Minneapolis, Minnesota?

Obstacles to Cultural Understanding

Some of the more serious obstacles to understanding other cultures are *ethnocentrism,* *prejudice* and *racism,* and *cultural relativity.*

Ethnocentrism

Pride in one's own culture, or **ethnocentrism,** can be both positive and negative. Such pride helps people bond, gives them a sense of unity, and can increase self-esteem. Carried too far, however, ethnocentrism can be risky. Many Americans, for example, view their culture as the "best" culture. They believe that any standards, practices, or behaviors that are different from theirs are inferior. They may belittle or ridicule them. They may even try to persuade others to change their ways. Americans are not alone in their ethnocentrism—people of almost any culture express it. Excessive ethnocentrism can lead to prejudice, stereotyping, and, ultimately, racism.

Prejudice and Racism

Prejudice operates on three levels: It affects your thinking, your emotions, and your behavior. When prejudice affects your thinking, you believe in stereotypes. In other words, you assign certain behaviors or characteristics to every member of a certain group; you do not look at each person as an individual. A stereotype is a conventional, formulaic, and oversimplified conception or image. These generalizations may be favorable or unfavorable, and they are often emotionally shaded. When prejudice operates at the emotional level, it affects your feelings toward another person. You generally feel superior to that person. At the behavioral level, you act on your prejudices by engaging in discriminatory behavior.

Everyone has some type of prejudice. It's a part of being human. Increased awareness of this is the first step toward becoming less prejudicial. Then you must recognize that you were not born with these prejudices—that you learn them through social institutions, such as the family, workplace, peer groups, and government.

While prejudice can probably never be eliminated, racism can be. Racism is an outward manifestation of prejudice that results in the unequal treatment of people so one group can gain a social advantage over another. Racist behaviors and beliefs are based on an ideology that justifies such treatment on the grounds that one race is inherently superior to another or that one is more deserving than another.

Cultural Relativity

Culture is relative. For example, the word *house* can be translated into a multitude of languages, but it conjures up very different images for people of different cultures. An American might imagine a two-story suburban home with a large yard and a white picket fence. But what about Asians or Africans? Is their concept of *house* the same as yours?

Or consider the cultural implications of the word *rain.* In Western cultural tradition, rain connotes God's wrath in the tale of the great flood and Noah's Ark. Images of thunder, lightning, and torrential downpours have come to symbolize danger and something ominous. Horror movies usually include a scene of a dark and stormy night. English-speaking people have incorporated this concept of rain into phrases such as "saving for a rainy day," "the eye of the storm," or "the calm before the storm."

But *rain* has quite a different cultural connotation among the native people of the American Southwest. For people dwelling in the desert, who depend on melting mountain snow as a primary source of water, rain is seen as a divine gift. Reflect on this Zuni prayer:

Cover my earth mother four times with many flowers.
Let the heavens be covered with the banked-up clouds.
Let the earth be covered with fog; cover the earth with rains.
Great waters, rains, cover the earth. Lightning cover the earth.
Let the thunder be heard over the earth; let thunder be heard.
Let thunder be heard over the six regions of the earth.

Zuni prayer

Critical Thinking

Would you treat water differently in the presence of people who live in a desert and prize it as a gift from the divine?

Culture and Military Operations

Cultural awareness helps identify points of friction within populations, helps build rapport, and reduces misunderstandings. It can improve your unit's ability to accomplish its mission and provide insight into individual and group intentions. As an Army leader, you must keep up to date on world events (be culturally astute) and be able to use this awareness and understanding to conduct operations innovatively—particularly in those areas where the United States has national interests. You may have to interact with people who live in those areas—as partners, neutrals, or adversaries. The more that you and your Soldiers know about these people's cultures, the better prepared you will be.

Failure to understand the cultural effect of an act may have severe repercussions. For example, in May 2005, the British press released photos of former Iraqi president Saddam Hussein in prison. Such photos are not unusual in British press. Other photos and coverage showed an American medical technician examining Saddam Hussein's teeth. Many Muslims, however, were offended by the pictures. They felt that the photos were deeply disrespectful of Islam. They held widespread anti-US demonstrations. This reaction complicated operations for US Soldiers and allies in the COE.

The following vignette explains some of the difficulties that American Soldiers have encountered as they deal with Iraqis on a daily basis. As you read, see if you can identify some of the areas of misunderstanding and think about how you might overcome them.

Baffled Occupiers, or the Missed Understandings

BAGHDAD, Iraq—Pollsters and journalists have been busy asking Iraqis how they feel about the Americans on their streets, but there is a potentially more important issue. How do the Americans here feel about the Iraqis? Will they ever feel comfortable enough in this alien culture to finish the job they started?

In public, the Americans have so far been the relentlessly chipper counterpoint to the chorus of complaining Iraqis. In private, though, you hear complaints like the one from a G.I. wearily minding a checkpoint one night in Baghdad looking for black-market gasoline, guns, and explosives. Asked how things were going, he shrugged disgustedly and said, "If you really want to know, I'm sick of being in a country where lying is the national pastime."

You can dismiss his assertion about Iraqis, which reflected his frustrations as well as his ignorance. But the remark reflects a cultural misunderstanding here that could ultimately matter more than the protests of Iraqis, the current violence against foreigners, or perhaps even billions of dollars for reconstruction.

Of course, security concerns are immensely important. They have already caused relief agencies to pull out most of their Western workers, leaving many of the remaining foreigners isolated in military and civilian compounds.

Lt. Travis Horner, a platoon leader from Phoenix, Ariz., stationed with the 82nd Airborne Division, said that he had been ambushed on five separate occasions, but that those were not the worst moments for him or his troops. The worst moment came when they learned they were stuck for an extended tour at their base in Habbaniya, a dust bowl 40 miles west of Baghdad.

"The funny thing is that our extension over here for a year is actually tougher on morale than enemy contact," Lieutenant Horner said.

The problems of physical isolation are compounded by the language barrier. Few American civilian and military "experts" on the Middle East speak Arabic (and even fewer read it, which is more difficult). Viewers of Al Jazeera are accustomed to a stark contrast between Russian diplomats speaking mellifluous classical Arabic and American diplomats speaking through a translator.

But those problems, while annoying, could be dealt with. There are deeper differences that may never be bridged.

When Americans do venture out of their compounds, they find that Iraqis are remarkably hospitable. "Welcome" is probably the most widely used English word here. Even if his kitchen has just been destroyed by a car bomb, an Iraqi host will apologize to a visitor for not offering the ritual cup of tea.

But Americans often do not know how to reciprocate politely. They routinely offend Iraqis by plunging bluntly into business instead of paying respects to the host and asking questions about his well-being and that of his family.

Iraqi culture often seems unfathomable to Americans because of the supreme importance attached to preserving personal and family honor. The code of honor makes it difficult to know what half the population is thinking, since men routinely shield their wives and daughters from contact with outsiders.

Even with the men, any topic can be taboo if there is a potential threat to someone's dignity, no matter how innocuous the issue seems to an outsider. Sometimes a translator will gently tell an inquisitive American, "You can't ask that question here," although the more common technique is for a translator to spare the American's feelings as well by simply asking a different question in Arabic.

These disjointed conversations can be maddening to American Soldiers and planners who cannot understand why any other value takes precedence over efficiency and "transparency," that great buzzword among Western political reformers.

"It's fairly common for an Iraqi to tell an American he understands something when he doesn't," said an Iraqi-American businessman with recent experience in Baghdad. "The American will name a material to use in building a wall, and the Iraqi will agree to do it even though he's never heard of it. He doesn't want the embarrassment of admitting he doesn't know."

Some American Soldiers complain that they have never heard such incredible face-saving stories. "You can catch a guy stinking of alcohol who's breaking into a liquor store in the middle of the night," one Soldier said, "and the next day he and his family will swear that he was just going in there to do some work for a friend, and that he would never drink because it's forbidden by the Koran. Family honor is the premium concern."

Pollsters working in the Middle East give interviewers special training to overcome respondents' tendency to guess the "right" answer instead of express their true feelings. That tendency is especially acute in Iraq after decades under Saddam Hussein, when a wrong reply could mean death.

One result is a wariness that discourages Americans, many of whom end up spending most of their time inside their compounds talking to one another while dispatching Iraqis into the field.

"It's a constrained existence we lead here," one occupation official said. "I wouldn't be here if I didn't think we could make a difference in Iraqis' lives, but it's awkward to interact with them because of the security concerns and other cultural barriers. This was a totalitarian country where people learned to survive by hiding their true feelings and fearing people in authority. Those habits don't just disappear."

A Soldier hands out toys to Iraqi children.

With time, perhaps, Americans and Iraqis can overcome these many barriers, but Americans may not have the patience. It's hard enough working in a dangerous, uncomfortable place far from home and family. It's tough enough hearing constant criticism from politicians and journalists. But the job can look terminally thankless if you cannot even understand the people you're trying to save.

The New York Times

The Leader's Role in an Organization

In this section, you've learned about what culture is, how it is transmitted, and why it is important to a society or people. Now it's time to relate these principles to your profession. What is your role in forming, maintaining, and changing organizational culture in the US Army? How do you, a leader, translate your understanding into results?

When an organization is created, leaders establish its initial rules, practices, customs, rituals, and methods. These become the espoused values of the group. Values include beliefs on such topics as toleration, stability, prosperity, social change, and self-determination. Each group a person belongs to instills in that person its values, with their ranking in importance. Individuals do not unquestioningly absorb all of the values of the groups they belong to; they accept some and reject others. But organization members constantly observe and take cues from their leaders. Leaders' questions, comments, reactions to events, emotional displays, and other behaviors convey strong messages to organization members and thereby influence the organizational culture. For these reasons, leaders are generally the primary source of the organization's initial culture.

Leaders influence the beliefs and behaviors that the organizational culture incorporates by what they *pay attention to, measure,* and *control.* These three factors indicate to other members of the organization what the leader believes is important. As a leader, you can exert a powerful effect on your unit's culture through systematically employing various culture-building techniques. Some are subtle; others are obvious. By establishing strong unit morale, standards, traditions, measures of performance, lines of communication, and a good work ethic, you can influence your Soldiers and create a culture that will help build a successful unit.

The Importance of Leader Understanding

It's important that leaders be aware of and understand culture. Understanding how organizational culture develops and acknowledging that alternative cultures can be equally valid will make you more alert to opportunities for organizational change and more willing to pursue such change. You will realize the realm of cultural possibilities, appreciate the importance of culture, and know what you must do to change it.

For example, the organization's shared basic assumptions may result in member behaviors that work against the achievement of the leader's vision and goals. Once you've identified these, you can work to change such counterproductive cultural elements. And, in the same way, you may build on and strengthen elements of the organizational culture that aid in accomplishing organizational vision and goals.

Understanding Your Own Unit's Culture

One of the most important cultures for you to understand is the culture of your own unit. Each unit, you will discover, has a culture of its own. That culture depends a good deal on the cultures of the individual Soldiers in the unit.

How do you go about creating a unit culture? The first step is to be sensitive to the backgrounds of your Soldiers. Know their histories and cultures. Your Soldiers may find it difficult to articulate their shared basic assumptions, but with sensitive and persistent questioning, you can help them express themselves clearly.

In other cases, you may assume command of a unit whose culture is already well established. In that case, listening and learning become all the more important. Your NCOs and subordinate leaders are a valuable resource: They are the repositories of the unit's "living memory." With patience, flexibility, and a deepening appreciation for how organizational culture works, you will be able to influence your unit's ongoing cultural development.

Critical Thinking

What are some of the effects of culture—both the Army's internal culture and the culture of other countries—on military operations?

CONCLUSION

Understanding other cultures is essential for all Army units and operations. Such an understanding could help you avoid Lord Chelmsford's error: For example, you would employ different tactics against an adversary who considers surrender a dishonor worse than death than you would against an enemy who sees surrender as an honorable option. Likewise, if your organization operates as part of a multinational team, the degree to which you understand your partners from other countries, and their understanding of you, will affect how well your team accomplishes its mission.

As an Army officer, you must take the time to learn the customs and traditions of your adversaries, as well as of the US Army, its units, and US allies. You must learn how and why others think and act as they do. This will help the United States outwit and outfight its foes and maintain its allies. Cultural awareness is critical to your professional success and the outcome of your mission.

Key Words

culture
artifacts
espoused values
shared basic assumptions
ethnocentrism

Learning Assessment

1. Describe Dr. Edgar Schein's three levels of organizational culture.

2. Explain ethnocentrism in both its positive and negative aspects.

3. Explain how cultural awareness can make you a better Army leader.

4. Explain how culture affects military units and operations.

References

Field Manual 3-0, *Operations*. 27 February 2008.

Field Manual 3-24, *Counterinsurgency*. 15 December 2006.

Field Manual 6-22, *Army Leadership: Competent, Confident, and Agile*. 12 October 2006.

Miraglia, E., Law, R., & Collins, P. (1999). *What Is Culture?* Washington State University. Retrieved 11 December 2005 from www.wsu.edu/gened/learn-modules/top_culture/culture-definition.html

Schein, E. H. (1992). *Organizational Culture and Leadership*. 2nd Edition. San Francisco: Jossey-Bass.

Sung, B. L. (1989). Bicultural Conflict. *The World and I*, August 1989. Retrieved 11 December 2005 from http://www.worldandi.com/specialreport/1989/august/Sa15247.htm

Tierney, J. (2003, October 22). Baffled Occupiers, or the Missed Understandings. *New York Times*, p. A4. Retrieved 20 October 2008 from http://query.nytimes.com/gst/fullpage.html?res=9B0CE3DD1F3EF931A15753C1A9659C8B63

Zuni Prayer. (trans. 1902). *In the Trail of the Wind*, (1971), ed. by John Beirhorst. New York: Farrar, Straus & Giroux.

CULTURE OF TERRORISM

Key Points

1 The Terrorist Mindset

2 Characteristics of Terrorists

3 The Terrorist Planning Cycle

4 Terrorist Operations, Tactics, and Weapons

The terrorists want to attack our country and harm our citizens. They believe that the world's democracies are weak, and that by killing innocent civilians they can break our will. They're mistaken. America will not retreat in the face of terrorists and murderers. And neither will the free world. As Prime Minister Blair said after the attacks in London, "Our determination to defend our values and our way of life is greater than their determination to cause death and destruction to innocent people." The attack in London was an attack on the civilized world. And the civilized world is united in its resolve: We will not yield. We will defend our freedom.

President George W. Bush
Remarks delivered at the FBI Academy

Introduction

While the late 20th and early 21st centuries have witnessed a dramatic rise in worldwide terrorist activity, terrorism itself is actually ancient—and ongoing. Terrorism is the calculated use of unlawful violence or the threat of unlawful violence to inculcate fear—terrorists intend to coerce or intimidate governments or societies in the pursuit of political, religious, or ideological goals. As long as there have been armies and political conflicts, there have been terrorists. Whenever fighters have deliberately stepped outside the bounds of traditional military tactics, particularly at the expense of civilian populations, the shadow of terrorism has loomed.

Terrorists are often portrayed as unpredictable, viciously irrational, and obsessed with media images and sensationalism. This is not the case. Terrorists rationally select their tactics and employ them in the pursuit of social, political, or religious aims. That said, some terrorists may not always be concerned with particular causes or ideologies. Some who use terrorism may be motivated purely by a need to be terrorists, in whatever cause suits them, or as a "hired gun" or mercenary serving a variety of causes for money.

The following news story reports several of the different techniques terrorists have employed to disrupt civil authorities and the orderly process of installing a democratic government in Iraq.

Suicide Bomber Attacks Shiite Mosque in Third Straight Day of Violence

BAGHDAD, Iraq—A leading Sunni cleric called for religious and ethnic groups to take a stand against violence as Iraq endured a third consecutive day of sectarian killings—the worst, a suicide car bombing at a Shiite mosque that killed at least 12 worshippers as they left Friday prayers.

The bombing in Tuz Khormato, where a young Saudi man was later arrested wearing a bomb belt on his way to a second mosque, was the latest suicide attack following al-Qaida in Iraq's declaration of all-out war on Iraq's Shiite Muslim majority.

Jordanian-born Abu Musab al-Zarqawi's terror group said it was taking revenge for a joint Iraqi–US offensive against its stronghold in Tal Afar, a city near the Syrian border.

With more than 20 people killed Friday, the death toll over the past three days surpassed 200, with more than 600 wounded.

Sheik Mahmud al-Sumaidaei, a leading Sunni cleric whose group is linked to the country's insurgency, criticized militants for targeting civilians. He called for Iraq's religious and ethnic groups to take a stand against further bloodshed.

"I call for a meeting . . . of all the country's religious and political leaders to take a stand against the bloodshed," al-Sumaidaei said during his sermon at Baghdad's Um al Qura Sunni mosque.

"We don't need others to come across the border and kill us in the name of defending us," he declared, a reference to foreign fighters who have joined the insurgency under the banner of al-Qaida. "We reject the killing of any Iraqi."

In Tuz Khormato, 130 miles north of Baghdad, authorities said the attacker detonated his explosives-packed car as worshippers flowed out of the Hussainiyat al-Rasoul al-Azam mosque, a Shiite Turkmen place of worship.

Police said 12 people were killed and 23 wounded in the bombing, which also destroyed 10 shops and eight cars.

"We were stepping out of the mosque and suddenly a big blast shook the ground," said Mustafa Ali, a 63-year-old ethnic Turkmen who escaped injury.

"I saw many people scattered on the ground, drenched in their own blood. I wanted to ask the bomber, 'Why did you attack those innocents who had prayed?'" he said.

Police Capt. Mohammed Ahmed said his men exchanged gunfire with another bomber before capturing him as he fled toward a second mosque. The man, who appeared to be in his early 20s, said he was from Saudi Arabia.

Friday's bloodshed began early, when gunmen opened fire on day laborers in an east Baghdad Shiite district. Three workers died and a dozen were wounded in the drive-by attack.

In Haswa, 30 miles south of Baghdad, a car bomb exploded near an Iraqi police patrol, killing three officers and wounding four, said police Capt. Muthana Khalid.

Gunmen also stormed the house of the mayor in nearby Iskandariyah, killing him and four bodyguards.

Sheik Fadil al-Lami, the Shiite cleric at Baghdad's Imam Ali mosque, was gunned down as he waited to gas up his car, said police Col. Shakir Wadi. Authorities also found the bodies of three people in the same Shiite district, including one Iraqi soldier.

Associated Press

The Terrorist Mindset

No one profile can describe all terrorists in terms of their backgrounds or personal characteristics. Although it is impossible to predict the identity and mindset of future terrorists, one can make some valid generalizations about terrorists today.

Utopian Worldview

Utopianism is a worldview that aims for a perfect society and ignores many practical realities of culture, politics, and tradition. Terrorists typically have *utopian* goals, regardless of whether their aims are political, social, territorial, or nationalistic. This utopianism expresses itself forcefully as an extreme degree of impatience with the rest of the world, an impatience that validates their extreme methods. Terrorists want the world to match their vision *now*. The individual terrorist commonly perceives the need for change to be so urgent that only the most extreme methods will succeed.

Interaction With Others

Terrorists interact in common ways with other members of their groups and with their leadership. Individuals who form or join terrorist groups commonly adopt the "leader principle," which requires unquestioning submission to the group's authority figure. Another adaptation the individual makes is accepting an us-against-the-world mentality. This results in viewing the acts of other individual members of the group as automatically moral, the cause as pure, and the goals as righteous.

Dehumanization of Nonmembers

The flip side of this interaction with their leader and fellow group members is that terrorists typically dehumanize everyone outside their group. This **dehumanization** allows them to indiscriminately target anyone outside the group, removing any guilt or remorse about killing innocent civilians—including women and children.

> **dehumanization**
>
> *to deprive of human qualities, such as individuality, compassion, or civility*

Lifestyle Attractions

Many terrorists find their lifestyle satisfying and purposeful. While not particularly appealing for members of stable societies, terrorists reap emotional, physical, and sometimes social rewards from being a member of a terrorist cell. They get a psychological "rush" and can feel a sense of elitism and a freedom from society's rules. Susan Stern, a member of the Weather Underground group in the United States in the early 1970s, described her involvement with the group: "Nothing in my life had ever been this exciting!"

Motivation

"[Al Qaeda (AQ)] and its loose confederation of affiliated movements remain the most immediate national security threat to the United States and a significant security challenge to the international community," the State Department says. "AQ still retains . . . the intent to mount large-scale spectacular attacks, including on the United States and other high-profile Western targets. . . . AQ's current approach focuses on propaganda warfare—using a combination of terrorist attacks, insurgency, media broadcasts, Internet-based propaganda, and subversion to undermine confidence and unity in Western populations and generate the false perception of a powerful worldwide movement." The radical Islamic militants and other terrorists that attack US forces are motivated by several factors:

US Presence

Many terrorists oppose the presence of US military forces in a particular area or the presence of organizations that US forces are safeguarding. For example, a number of terrorist groups strongly object to any US presence in Saudi Arabia because of the groups' ties to fundamentalist Islam. This presence of foreign "infidels" on what they see as holy ground may be reason enough in their minds to attack.

Culture

Antagonists who directly oppose one or more major characteristics of American culture—such as capitalism, secular democracy, racial tolerance, or the equality and role of women—may attack Americans wherever they are found throughout the world.

State of Conflict

Groups that feel they are "at war" or in a social or political conflict with the United States will target military personnel and facilities to gain legitimacy and "make a statement." Likewise, nation-states at war with, or about to go to war with, the United States will use terrorist organizations, clandestine military operations, or intelligence assets to attack US military targets.

Critical Thinking

While terrorists are a nontraditional, highly versatile enemy, which traditional methods and tactics of warfare might be effective in neutralizing them?

Characteristics of Terrorists

Although no single personality profile of a terrorist exists, some general characteristics are fairly common.

Status

Terrorists rarely arise from a background of poverty and despair. They are much more likely to spring from middle class backgrounds, with some even coming from the upper class of their society. You must understand these terms in the context of the terrorists' society of origin, however: "Middle class" or "privileged" are relative terms and will represent completely different levels of income in West Africa and Western Europe, for example.

Education and Intellect

Left-wing terrorists, the leadership of right-wing terrorist groups, and terrorists of either type who operate internationally usually have average or higher-than-average intelligence and at least some advanced education. (Osama bin Laden and Yasir Arafat were both educated as civil engineers.) While many terrorists generally have had exposure to higher learning, they are often not highly intellectual and are frequently dropouts or possess poor academic records. This, again, depends on the norms of their society of origin. In societies where religious fundamentalism is prevalent, higher education might mean advanced religious training.

US domestic and right-wing terrorists tend to come from lower educational and social levels, although they have some education. Right-wing domestic groups in the United States were quick to explore the Internet's communications and organizational potential. Such terrorists will typically have a high school education and be very well indoctrinated in the ideological arguments they support.

Age

Terrorists tend to be young. Leadership, support, and training cadre can range into their 40s, but most operational members of terrorist organizations are 20 to 35 years old. Individuals under 20 usually lack the practical experience and training necessary to be effective. Some groups, however, recruit children to offset personnel shortages and "grow" future terrorists.

Gender

Terrorists are not exclusively male, even in groups that are rigorously Islamic. Women's roles in these groups will often be limited to support or intelligence work, but some fundamentalist Islamic groups use women in operational roles, including suicide missions.

Appearance

Terrorists do not normally appear out of the ordinary and are capable of normal social behavior and appearance. Over the long term, elements of fanatical behavior or ruthlessness may become evident, but it is typically not immediately obvious. Although members of so-called sleeper cells or other covert operators may marry as part of their disguise, most terrorists do not marry—although notable cases of married couples within terrorist organizations have been reported.

The Terrorist Planning Cycle

Terrorists usually meticulously prepare their operations to minimize risk and achieve the highest probability of success. They focus on avoiding their opponents' strengths and concentrating on the opponents' weaknesses. They work to achieve as much security and as powerful an effect on their target as possible. In practice, that means they use the fewest number of personnel and the most effective weapons available. To accomplish this, terrorists plan extensively, with an emphasis on target surveillance and reconnaissance.

There is no universal "staff school" model for terrorist planning. Experience and successes, however, have shown terrorists what works best for effective planning and operations. Their planning generally goes through seven basic phases:

Phase I: Broad Target Selection

This phase involves collecting information on a large number of potential targets, some of which the terrorists may attack or seriously consider attacking. The personnel who collect this information are typically not core members of the terrorist organization, but are either sympathizers or dupes who may not even be aware of the purpose of their work. This phase also includes collecting open-source and general information, much of it available through the Internet.

Phase II: Intelligence Gathering and Surveillance

The terrorists then focus their efforts on targets they believe to be most vulnerable. This focus leads to additional information gathering on the targets' patterns over time. The type of surveillance the terrorists use depends on the target. Information gathered covers people's routines, physical layouts, routes of travel, and security measures surrounding the target.

Phase III: Specific Target Selection

In selecting a target for actual operational planning, terrorists consider some of the following factors:

- Does success affect a larger audience than the immediate victim(s)?
- Will the target attract high-profile media attention?
- Does success make the desired statement to the correct target audience(s)?
- Is the effect consistent with the group's objectives?
- Does the target provide an advantage to the group by demonstrating its capabilities?
- What are the costs versus the benefits of conducting the operation?

A decision to proceed requires continued intelligence collection against the chosen target.

Phase IV: Pre-Attack Surveillance and Planning

At this point, members of the actual operational cells begin to show up. They are either trained intelligence and surveillance personnel or members of the cell organized to conduct the operation. During this phase, the terrorists gather further information on the target's current patterns over time, usually days to weeks. This allows the attack team to confirm the information gathered from previous surveillance and reconnaissance activities. The areas of concern are essentially the same as in Phase II, but with greater focus on the specific planning conducted thus far.

Phase V: Rehearsals

As with conventional military operations, rehearsals improve the odds of success, confirm planning assumptions, and allow members to develop contingencies. Terrorists also rehearse to probe and test security reactions to particular attack profiles. Terrorists use both their own operatives and unwitting people to test target reactions.

Phase VI: Actions on the Objective

Once terrorists reach this stage of their program, the odds are clearly against the target. Several different analyses have concluded that once the terrorists initiate their operations, they succeed 90 percent of the time. Terrorists will spend as little time as possible conducting the actual operation in order to reduce their vulnerability to discovery or countermeasures. With the exception of barricade-style hostage-taking operations, terrorists plan to complete their actions before nearby security forces can react.

Phase VII: Escape and Exploitation

Most terrorists want to survive the operation and escape. Their escape plans are usually well thought out and executed. Escape further enhances the effect of fear and terror from a successful operation if the perpetrators get away "clean." The exception to this is suicide operations, where the impact is enhanced by the attacker's apparent willingness to die. Even in suicide attacks, however, support personnel and handlers usually deliver the suicide bomber to the target and subsequently make their escape.

Critical Thinking

Can you analyze terrorists' actions and planning using the characteristics of the offense and the principles of war?

Terrorist Operations, Tactics, and Weapons

Terrorists tailor their tactics, forces, and weapons specifically to the particular mission. Each terrorist operation is planned for a specific target and effect. Additionally, terrorists will expose only as much of their resources and personnel to capture or destruction as absolutely necessary to complete their mission.

Understanding the objectives of the group conducting the operation is key to predicting likely targets. Although several different types of operations may satisfy a particular objective, terror groups often develop expertise in one or more types of operation and less specialization in others.

Typical terrorist operations include:

- Assassination
- Hostage taking and barricade situations
- Kidnapping
- Raids
- Extortion
- Ambush
- Hijacking
- Sabotage
- Aircraft attacks
- Attacks on ships.

Terrorists use a number of tactics and techniques to accomplish these operations, including bombing; arson; hoaxes, misdirection, and multiple attacks; and suicide tactics.

Weapons and Equipment

Analyzing the weaponry and equipment available to a terrorist group is an important part of any assessment of organizations that use violence. Whereas conventional military organizations rely upon standardization, terrorists rely upon weapons and equipment tailored to each new operational requirement. If a 30-year-old RPG-7 will do the job, they will use it. If not, the terrorists will buy the weapons they need. Since terrorists, unlike conventional armed forces, do not have to go through long and bureaucratic acquisition processes, their only limitations in obtaining state-of-the-art systems are financing and availability of the equipment. If a sophisticated precision guided missile is needed and they cannot buy it, they will "build" it, using a suicide bomber and the appropriate explosives.

Terrorists have been able to obtain extremely dangerous weaponry due to state sponsorship of many terrorist groups, arms flows to regional conflicts, and a widespread illegal international arms trade. This is one reason that detecting and interrupting terrorist groups' movement of money through financial institutions worldwide is a crucial tactic against terrorism.

Firearms

Terrorists use a variety of firearms, including submachine guns, pistols, assault rifles, sniper rifles, and shotguns. When selecting weapons, terrorists look for three major characteristics: availability, simplicity, and efficiency. They like automatic weapons that can kill from a distance and have stopping power. They also want to be able to conceal the weapon, especially in urban terrain.

Terrorists do try to standardize calibers of their weapons as much as possible to ease ammunition resupply. They favor easily available military and semi-military weapons. Most international terrorist groups favor fully automatic weapons, such as the AK-47 and the M16. A weapon favored by small groups in the United States is the 12-gauge shotgun.

Pistols

Pistols are standard weapons for terrorists. They are small and easily concealed. Most of them are lightweight, and many modern pistols provide good firepower. Since their effective range is generally limited to about 50 meters, they require the attacker to be fairly near the target. They can be very effective at close range.

Submachine Guns

Submachine guns have a full automatic fire capability, use pistol-caliber ammunition, and typically have large magazine capacities. Their range, accuracy, and penetration are better than that of pistols because of their longer barrel and sight radius. Submachine guns are a favorite with terrorist groups because they are small, light, and easily concealed. They provide a large amount of firepower and are deadly at close range.

Assault Rifles

Assault rifles are the primary offensive weapons of modern militaries and used extensively by terrorist organizations. They normally have selective firing capability to allow single shot, two- or three-round bursts, or full automatic mode. Their effective ranges can often exceed 600 meters, and they have effective rates of fire up to 400 rounds per minute in full automatic mode. They provide terrorists the same firepower as they do for a modern Soldier on the battlefield.

Sniper Rifles

Since one of the primary terror tactics is assassination, terrorists often use sniper rifles to attack targets that are difficult for other weapons to reach. With the development of large-caliber sniper weapons, terrorists can also effectively engage light-armored vehicles.

Shotguns

Although limited in range and penetration capability, shotguns are excellent close-range weapons, especially for assassinations. Shotguns require less precision in aiming since the dispersion of buckshot allows a large number of pellets to cover a wide area. They are readily available and relatively inexpensive compared with other weapons. Additionally, the barrels can be sawed off to permit easy concealment and increased dispersion of shot.

Munitions

Terrorists have used grenades in all types of attacks, including rocket-propelled grenades (RPGs). Although originally developed as an antitank weapon, RPGs double as terrorist antiaircraft weapons. Attackers brought down two MH-47 Chinook helicopters in Afghanistan in 2002 using RPGs. In 1993, a pair of UH-60 Black Hawks suffered a similar fate in Mogadishu, Somalia.

Terrorists also favor portable surface-to-air missiles (SAM). These weapons proved to be highly effective in the hands of the Afghan Mujahideen guerrillas during their insurgency against the Soviets in the 1980s. In a number of documented cases, terrorists have used them against civilian airliners.

Although terrorists use mostly "homemade" bombs, they do use some conventional munitions, especially as booby traps. They often obtain unexploded ordnance and modify it for their purposes. Terrorists have used various forms of aerial bombs, as well as artillery and mortar rounds, in this fashion. Additionally, they employ conventional mines to engage a variety of targets.

The following are examples of common explosive charges terrorists have employed to great effect and success:

- Improvised explosives, such as fertilizer, black powder, gasoline, match heads, and smokeless powder
- Chemical reactions, such as acid bombs; caustic bombs, such as Drano and dry ice
- Plastic explosives, such as C-4 and SEMTEX
- TNT
- Dynamite.

These explosive devices require some kind of trigger to detonate them. Triggers range from very simple homemade devices to highly technical ones. Some examples are:

- Manual wind-up alarm clocks and wristwatches
- Pressure release switches, such as in mousetraps
- Pull switches that activate with a tripwire
- Pressure switches
- Wire-command detonation
- Radar guns.

Improvised Explosive Devices (IEDs)

improvised explosive device (IED)

a "homemade" ordnance used to disable or destroy personnel or vehicles from a hidden emplacement

While terrorists will use conventional weapons, such as RPGs and assault rifles, to achieve their tactical goals, they also assemble and employ a wide variety of lethal **improvised explosive devices (IEDs)**. They incorporate highly destructive lethal and dangerous explosives or incendiary chemicals designed to kill or destroy the target. They often steal the materials they need for these devices from military or commercial blasting supplies. If necessary, they will also use fertilizer and other readily available household ingredients.

IEDs basically include some type of explosive, a fuse, detonators and wires, shrapnel and pieces of metal, and a container in which to pack the explosives and shrapnel. Although terrorists use manufactured explosive material, they can easily obtain the ingredients required to make improvised explosive material as well.

The types of IED vary based on the explosive used, method of assembly, and the method of detonation. Terrorists can use infinite combinations to manufacture them. Some common IEDs are:

- Pipe bombs
- Incendiary devices, such as Molotov cocktails
- Vehicle devices, often known as a vehicle-borne IED (VBIED)
- Projected IEDs, such as platter charges and improvised mortars.

New Threats: Weapons of Mass Destruction (WMDs)

Of greatest concern to governments fighting global terrorism are the relatively recent efforts by terrorists to acquire biological, chemical, and nuclear materials to make **weapons of mass destruction (WMDs)**. Today these weapons present the greatest threat to large populations in areas targeted by terrorists. Terrorists in the Contemporary Operating Environment (COE) are actively seeking WMDs that they can easily conceal, transport, and covertly deliver without warning.

weapons of mass destruction (WMDs)

nuclear, biological, and chemical weapons designed to kill large numbers of people

CONCLUSION

As an Army leader in the COE, you will need to be ever more vigilant in combating the forces of global terrorism. Your platoon may be a primary target for terrorists or insurgents employing terrorist tactics. With the stakes no less than US values and the American way of life hanging in the balance, you will need to hone your counterterrorism expertise and train your Soldiers to be alert to and defeat the threat wherever it operates.

Remember, terrorism is not a random, mindless activity. It consists of operations carefully planned by a wily, calculating, and deadly enemy. US forces—starting with you and your Soldiers—must understand this enemy in order to disrupt terrorists' plans, interfere with their calculations, and neutralize them before they can execute their operations.

Key Words

dehumanization
improvised explosive device (IED)
weapons of mass destruction (WMDs)

Learning Assessment

1. Describe some of terrorism's characteristics.

2. List the elements of a typical terrorist profile.

3. Explain the operational phases of a terrorist action.

4. Name five types of terrorist operations.

5. Describe an improvised explosive device (IED).

References

Bush, G. W. (11 July 2005). Remarks delivered at the FBI Academy, Quantico, VA. Retrieved 5 December 2005 from http://www.whitehouse.gov/news/releases/2005/07/200507111.html

Counterterrorism. (n. d.). Federal Bureau of Investigation. Retrieved 5 December 2005 from http://www.fbi.gov/terrorinfo/counterterrorism/waronterrorhome.htm

Country Reports on Terrorism, 2006. (2007). Department of State. Retrieved 20 October 2008 from http://www.state.gov/s/ct/rls/crt/2006/82727.htm

El-Tablawy, T. (2005, September 16). Leading Sunni Cleric Calls for Halt to Violence. *Associated Press*. Retrieved 5 December 2005 from http://aolsvc.news.aol.com/news/article.adp?id=20050910141709990010. Used with permission of the Associated Press. Copyright (c) 2005. All rights reserved.

Field Manual 7-100.1, *Opposing Force Operations*. 27 December 2004.

Timeline of Terrorism. (2004). US Army. Retrieved 5 December 2005 from http://www.army.mil/terrorism/

NONGOVERNMENTAL ORGANIZATIONS, CIVILIANS ON THE BATTLEFIELD, AND HOST-NATION SUPPORT

Key Points

1 **Interacting With Other Organizations**

2 **Civilian Considerations and Military Operations**

3 **US Force Responsibilities**

4 **The Culture Maze**

Collaboration with NGO and international organizations is essential. The NGO and international organization community attempt to work together through consultation, coordination, consensus, and cooperation.

Field Manual 3-07.31, *Peace Operations: Multi-Service Tactics, Techniques, and Procedures for Conducting Peace Operations*

Introduction

The problems associated with noncombatant individuals and organizations on the battlefield are not new, as the following vignettes show. Note that in the account of Mrs. Judith Henry, Union Artillery CPT James B. Ricketts had to testify before a congressional committee in 1862.

Civilians on the Battlefield—Not a Modern Problem

Crowds

On a warm July day in 1861, two armies of a divided nation clashed on the fields overlooking Bull Run, a small stream near the little town of Manassas, Va. So confident that the northern army under the command of Gen. Irvin McDowell would be victorious, the wealthy elite of nearby Washington packed picnic baskets and headed to Manassas to watch the battle. Enthusiastic volunteers in colorful uniforms gathered to fight the first major land battle of the war. Confident that their foes would turn and run, neither side anticipated the smoke, din and death of battle.

The [First] Battle of Bull Run, also known as the [First] Battle of Manassas, was a rude awakening for Americans. Soldiers, mostly raw recruits, were stunned by the horror of battle. When the Union army was driven back in disorder, panicked civilians attempting to flee in their carriages blocked the roads to Washington.

Creators Syndicate, Inc.

Individuals

On Sunday, July 21, 1861, Mrs. Judith Henry, her daughter Ellen, and hired . . . girl, Lucy Griffith, were living at Spring Hill Farm. . . . When the battle of that day began on the opposite hill across Young's Branch [near Manassas], shots from the cannonading were coming threateningly near, the family first considered trying to get Mrs. Henry, who was bedridden from the infirmities of age, with soldier help, removed . . . ; but in the growing confusion this was out of the question.

When [CPT James B.] Ricketts' battery shelled the house, as he himself testified before a Congressional Committee the following year, to drive out the Confederate sharpshooters, the bed on which Mrs. Henry lay was shattered, she was thrown to the floor, being wounded in neck, side, and one foot partly blown off. She died later in the afternoon or early evening. Ellen Henry sought refuge in the big chimney to the fireplace during the bombardment and her subsequent deafness was attributed to injury to her eardrums from the violent concussion produced by the shelling. Whether John [Henry] was in the house during the shelling or not was never stated, but since he was unhurt, it is presumed that he was outside when the bombarding began.

Many years after the events of the day, an old man visiting the battlefield [said] that he was walking through the yard sometime after the close of the battle noting the many dead who had fallen fighting around the house when he came to a man lying face downward; and as he came up to this man, the man raised his face and said "They've killed my mother."

Unpublished manuscript in the files of Manassas National Battlefield Park

Conflict today is localized, asymmetrical, protracted, and multifaceted. Rather than capturing terrain, the purpose of today's warfare is to capture the hearts, minds, sentiments, and cooperation of the indigenous population. The aim is to change the political, and, at times, social infrastructure of the nation or region with which the US is in conflict. By simple acts of kindness or ignorance, you can make local nationals, individuals, and groups either assets or liabilities.

Today's battlefield contains not only combatants, but also **international organizations (IOs), nongovernmental organizations (NGOs),** displaced civilians, and refugees. You must account for these noncombatants when planning and conducting military operations. Their responsibility is to make themselves recognizable to combatants. Your responsibility, and that of your Soldiers, is to verify the status of individuals and groups to prevent them from taking aggressive actions. While this task is difficult, it's not impossible. When you include noncombatants in your planning considerations, you give yourself the opportunity to turn them into assets.

Military operations are more than just combat operations. As platoon leader, you will deal with enemy fighters and innocent civilians. While some combat operations continue, such as battling insurgents and terrorists, you may be engaged in support, stability, and peace activities. Some of these activities will require you to collaborate with NGOs, IOs, host-nation agencies, and other nonmilitary agencies to provide humanitarian relief to civilians who have been displaced by the fighting or by natural disasters such as hurricanes or floods. Such noncombat military operations may include:

- Processing and return of enemy prisoners of war (EPWs)
- Return of displaced civilians
- Evacuation of friendly civilians
- Transfer of responsibilities to follow-on, peacekeeping, or host-nation forces
- Restoring basic services, such as water, electricity, and health care.

To be successful in these and other operations, you must know how to interact with other elements within the operating environment. You must include them in your planning and, if possible, in your training. Cooperating with NGOs and other agencies can be a significant help in accomplishing your mission.

The following vignette illustrates the positive effects of cooperation during Operation Provide Comfort, an effort to provide humanitarian support to the Kurds in northern Iraq following Operation Desert Storm.

international organizations (IOs)

organizations with global mandates, generally funded by contributions from national governments— examples include the Red Cross, the World Food Programme, or the United National International Children's Emergency Fund (UNICEF)

nongovernmental organizations (NGOs)

transnational organizations of private citizens that maintain a consultative status with the Economic and Social Council of the United Nations— nongovernmental organizations may be professional associations, foundations, multinational businesses, or simply groups with a common interest in humanitarian assistance activities (development and relief)

Operation Provide Comfort

Following Desert Storm, the Kurdish population of Iraq attempted to flee the country to the north out of fear that Saddam Hussein would attempt to exterminate their entire population. Because of political concerns, Turkish officials refused to allow these desperate people permission to cross the border into Turkey. The result was that hundreds of thousands of Kurds were essentially trapped on barren and rocky hillsides, vulnerable not only to Hussein's forces, but also to the harsh elements. Without basic necessities, including access to water, food, and medical supplies, hundreds of Kurds were dying each week. In April of 1991, President George Bush made the decision to provide relief and protection for these beleaguered people. Literally overnight, Operation "Provide Comfort" was born. In less than 48 hours from receiving the order to "do something," cargo and fighter aircraft were re-deployed to bases in southern Turkey where they began delivering humanitarian supplies. Over a period of a few weeks, a US-led coalition force was deployed into northern Iraq, resettlement areas constructed, and a de-militarized zone established for the protection of the Kurds. . . .

Maj. Gen. James L. Jamerson . . . redesignated the organization as a Combined Task Force. The task force dropped its first supplies to Kurdish refugees on 7 April. As a result of President Bush's order and UN resolution 688, a coalition of 13 nations formed with material contributions from 30 countries working under the command and control of the Coalition Task Force. Although many nations ultimately contributed to the operation, the primary countries involved [in the mission] were the US, the United Kingdom, France, and Turkey. . . .

An important part of this mission was the "seamless" transfer of responsibility over to NGOs. Task Force Encourage Hope (later renamed Joint Task Force Bravo) was formed to construct a series of resettlement camps where dislocated civilians could find food and shelter and a secure environment. Encourage Hope was designed to integrate civilian relief agencies into the support, organization, and administration of the camps. The Kurds were expected to assist in the planning, construction, administration, and sustainment of these camps. The camps each held about 25,000 people and were initially supplied by the military. They eventually became self-sustaining and were transferred to Kurdish or non-government agency control as soon as possible. . . .

Task force members on the ground built refugee camps and maintained a security zone in northern Iraq to protect the Kurds from the Iraqi military. Air units operating from Incirlik enforced a no-fly zone above the 36th parallel while providing air cover for friendly forces on the ground. Aircraft from Incirlik and other bases in eastern Turkey dropped desperately needed supplies to the Kurds. . . .

Operation Provide Comfort (OPC) sought the achievement of two goals: To provide relief to the refugees, and to enforce the security of the refugees and the humanitarian effort. These two goals were maintained from April to September 1991 by the CTF. During this time it flew over 40,000 sorties, relocated over [700,000] refugees, and restored 70–80 percent [of the] villages destroyed by the Iraqis. . . .

By mid-July . . . the UN had assumed responsibility for the refugee camps.

GlobalSecurity.org

Critical Thinking

In Operation Provide Comfort, why was it essential to involve the Kurds as much as possible in setting up and running the resettlement camps? Why was it important to get civilian relief agencies and the UN involved as much as possible? What effect can including indigenous organizations and personnel have on mission accomplishment?

Interacting With Other Organizations

In just about any environment, you and your Soldiers will interact with other military services, agencies of other governments, the host or hostile nation, United Nations (UN) agencies, NGOs, IOs, humanitarian relief organizations, private voluntary organizations, and multinational partners, as well as with civilians.

You will also interact with US government agencies, usually including the Department of State, the United States Agency for International Development (USAID), and often the Department of Justice.

These organizations and agencies can play significant roles in helping you accomplish your mission. To interact with them effectively, you must understand their objectives and be familiar with the services that they can provide. You must also have the flexibility to interact with them cooperatively.

Nongovernmental Organizations

NGOs are nonprofit, voluntary citizens' groups that are organized on a local, national, or international level. They may be based in the United States or in another country. NGOs are task-oriented and driven by people with a common interest. They perform a variety of service and humanitarian functions, bring citizen concerns to governments, advocate, and monitor policy implementation.

Some NGOs are organized around a specific issue, such as human rights, the environment, or health care. Some NGOs help monitor and implement international agreements. NGOs may also provide early-warning mechanisms—for example, alerting authorities or world public opinion to a developing famine.

As you recall from your earlier study of stability and support operations, in noncombat situations the Department of State is generally the lead US government agency in a foreign country and must approve any collaboration with NGOs.

The number of NGOs involved in a particular support or stability operation varies. It is sometimes quite large. The US Agency for International Development (USAID) maintains a registration list of approximately 350 agencies capable of conducting some form of humanitarian relief operations. USAID publishes a yearly report titled *Voluntary Foreign Aid Programs* that describes the aims and objectives of its registered NGOs. Army forces frequently cooperate with large, well-known NGOs, such as the American Red Cross and World Emergency Relief, as well as with many smaller ones.

Soldiers who work with NGO staff need to be flexible and adaptable. In some cases, these organizations may initially hesitate to cooperate with military forces. Building mutual trust may take time. Cultivating good relationships is worthwhile, however, for most NGOs can provide useful information and insights concerning the local population. You should cooperate with NGOs as much as the mission allows. Cooperation and coordination with NGOs reinforce the legitimacy of US forces involved in a unified action.

The involvement of NGOs has many practical advantages. NGOs can dramatically reduce the military resources required for civil-military operations. NGOs have local contacts and experience. They are involved in such areas as education, technical projects, relief activities, refugee relief, public policy, and economic development. NGOs are frequently on the scene of a crisis before US forces arrive. They routinely operate in high-risk areas and usually remain at the site long after military forces have departed. For reasons such as these, Army leaders must integrate NGOs into planning, preparing, executing, and assessing military operations.

International Organizations

Unlike NGOs, which may have a limited sphere of activity, IOs have global influence. Examples include the International Committee of the Red Cross, the International Organization for Migration, and UN agencies, such as the UN High Commissioner for Refugees or the World Food Programme. IOs are well respected and can take on major responsibility—the International Red Cross, for example, is responsible for monitoring compliance with the Geneva Conventions.

Army Organizations

Helping a country in crisis requires skilled leaders who can promote good relations between US military and government organizations, NGOs, and IOs. The Army has a number of organizations that can help you carry out this work.

For example, if you need information on the implications of carrying out a certain mission, you may receive assistance from the unit G-5 office, the staff directorate that deals with civil affairs. Or you can request assistance from Army Civil Affairs personnel. Regionally oriented, language-qualified, and culturally attuned Civil Affairs forces can assist with training your forces and in dealing with various local issues you may face in your operations.

Civil Affairs personnel coordinate military operations with civilian agencies of the US government, civilian agencies of other governments, and NGOs. At higher levels, **civil-military operations centers (CMOCs)** accomplish this task. A CMOC coordinates the receipt and handling of requests for US military assistance from government organizations, IOs, and NGOs. CMOCs include representatives from many government agencies, and they call on the services of these agencies as needed for a particular mission.

Other Army organizations sometimes involved are public affairs, military police, and **psychological operations (Psyops)** units.

Host-Nation Support

During mission analysis and planning, you will need to pay special attention to the Contemporary Operating Environment (COE) and the host nation. Host nations can provide many kinds of support for US forces. Such support may include material resources, services, civilian labor, local security, and police forces.

Host-nation support is the civil and military assistance that a nation offers to foreign forces within its territory during peacetime, crises, emergencies, or war. That support is based on formal agreements among nations. Host nations provide assistance to meet the day-to-day needs of conducting military operations in a foreign country. They provide in-theater as well as en route support.

civil-military operations center (CMOC)

an ad hoc organization, normally established by the geographic combatant commander or subordinate joint force commander, to assist in the coordination of activities of engaged military forces as well as of other US government agencies, nongovernmental organizations, and regional and international organizations—its size and composition depend on the situation

psychological operations (Psyops)

planned operations to convey selected information and indicators to foreign audiences to influence their emotions, motives, objectives, reasoning, and ultimately the behavior of foreign governments, organizations, groups, and individuals

Other examples of host-nation support include:

- operation, maintenance, and security of seaports and airports
- construction and management of routes, railways, and inland waterways
- transportation support
- limited health services
- subsistence support
- laundry and bath support
- petroleum support
- bulk storage or warehouse space
- augmentation of communication and automation networks
- support of indigenous religious leaders.

Host-nation support agreements can significantly reduce logistics, maintenance, and personnel requirements for US combat service support forces.

Three types of groups are often involved in providing host-nation support. By becoming familiar with each type of group and the particular services it offers, you can ensure that services are of the highest quality and that resources are used efficiently.

Government agency support. Local government agencies build, operate, and maintain facilities and systems that can support US requirements. Examples of such systems include utilities and telephone networks. Police, emergency services, and border patrols may also be available to support US forces.

Civilian contractors. Local, national, third-country, or US contractors employing indigenous or third-country personnel in the host nation may provide supplies and transportation, labor, construction, and other services.

Local civilians. The US can turn to local workers to fill many roles, from laborers, stevedores, truck drivers, and supply handlers to highly skilled equipment operators, mechanics, computer operators, and managers. Individuals with these skills may be available in the host nation's labor pool.

As a leader, you should consider how all the services these groups provide can contribute to achieving your goals. When you work with them, you forge a vital link between military operations and the activities of NGOs, IOs, and governmental agencies of the US, host nations, and partner nations.

For additional information on some major NGOs and IOs, see one or more of the following:

United National High Commissioner for Refugees
(www.unhcr.ch)

Red Cross
(www.redcross.org)

Oxfam International
(www.oxfam.org)

World Vision International
(www.wvi.org)

International Medical Corps
(www.imcworldwide.org)

Save the Children
(www.savethechildren.org)

Critical Thinking

What are some positive aspects of using host nation support? What are some negative aspects?

Civilian Considerations and Military Operations

Civilian considerations (the *C* in METT-TC) generally focus on the immediate effect of civilians on the current operation. These considerations also include larger long-term diplomatic, economic, and informational issues, however. Still other civilian considerations include enemy locations with respect to civilian populations, political and cultural boundaries, and language requirements. Civilian considerations may preclude attacking certain targets, such as infrastructure and historically significant areas.

Controlling civilians is essential during military operations. For example, international law requires you to segregate civilians from enemy prisoners of war (EPWs) and civilian internees (CIs). Uncontrolled masses of people, friendly or not, can seriously hinder the military mission. According to US policy, the civilian host-nation government is responsible for the welfare of its population, including displaced civilians.

All commanders are subject to obligations imposed by international law, including the Geneva Conventions of 1949. In accordance with such law, commanders must establish law and order, protect private property within their geographic areas of responsibility, and provide a minimum standard of humane care and treatment for all civilians. You can get additional information from FM 27-10 and the staff judge advocate (SJA).

Media Coverage

Media coverage of operations in a host nation may affect civilians' attitudes toward you, your Soldiers, and the United States. Critical reports in US or other media—or enemy propaganda—may have a negative effect on attitudes and actions toward US Soldiers. A positive report, by contrast, may bolster civilian support. Such coverage may also affect US domestic and foreign support for an operation on a broad scale. Soldiers must pay particular attention to the effects of their actions in the information environment. Media satellite trucks can beam video and audio of Soldiers' actions worldwide in minutes. As an Army leader, you must monitor public opinion and keep your subordinates informed.

Population Movement

Combat operations often lead to a widespread displacement of civilians, a process known as *population movement*. Such movement never occurs without reason. It becomes likely only when the population believes that its rights—or often, its lives—may be in danger.

NGOs, IOs, and other local and national agencies can have key roles in controlling and overseeing civilians involved in population movement. Their involvement is essential for three reasons:

1. It conserves military resources
2. Civilian authorities normally have legal status and are best equipped to handle the needs of their own people
3. The use of local personnel reduces the need for interpreters or translators.

In this kind of situation, the US forces' role is to coordinate their efforts with those of these agencies.

Actions during population movement generally pass through five stages, each linked to a major operational concern or planning effort:

1. *Preflight and flight*—to provide intelligence support in determining the timing, size, and direction of a population movement

2. *Arrival*—to assist IOs, NGOs, and the host nation during the arrival of the refugees

3. *Asylum*—to build and maintain refugee camps and settlements while assisting with stabilization of the refugees' country of origin

4. *Repatriation*—to secure repatriation crossing points, screening points, transit sites, and movement of returnees to their communities

5. *Reintegration*—to help ensure the security of displaced civilians as they return to their communities. This support is especially critical if the host nation's public-safety authorities are overwhelmed or in the face of active resistance (from other ethnic or religious groups, for example) to resettlement. International civilian police normally assume the primary responsibility for community law and order in such circumstances.

Civilians Affected by Military Operations

During military operations, US forces must consider two categories of civilians: those who remain in place and those who are dislocated.

Those who remain. This category includes civilians who are indigenous to the area as well as individuals from other countries. The civilians in this category may or may not need help. If they can take care of themselves, they should remain in place.

Dislocated civilians (DCs). DC is a broad term for people who have left their homes or countries of origin for various reasons. Their movement and presence may hinder military operations. Most of them require some type of aid, such as medicine, food, shelter, and clothing. There are five categories of DCs:

1. Displaced persons—civilians who are involuntarily outside their area or region but still within their own country

2. Refugees—civilians who have left their homes to seek safety outside their countries because of real or imagined danger

3. Evacuees—civilians removed from their places of residence by local or national military order

4. Stateless persons—civilians who have been stripped of their citizenship (not necessarily legally), whose countries of origin cannot be determined, or who cannot establish their right to the nationality that they claim

5. War victims—civilians suffering injuries, loss of a family member, or damage to or destruction of their homes as a result of war. War victims may be eligible for a claim against the United States under the Foreign Claims Act. This category was created during the Vietnam era.

Soldiers questioning Iraqi civilians

Interacting With Displaced Civilians

The primary purposes of operations involving DCs are to:

- Minimize civilian interference in combat operations
- Protect civilians from combat operations
- Prevent and control the outbreak of disease
- Relieve, as far as is practical, human suffering
- Group DCs together as much as possible.

The State Department, the UN Office for the Coordination of Humanitarian Assistance (UNOCHA), and foreign civilian and military authorities should coordinate efforts to determine the appropriate levels and types of aid required and available for DCs.

For you and your Soldiers, successful interaction with DCs, as well as with local citizens who have remained in place, requires discipline. In circumstances where the populace is ambivalent or unfriendly, discipline fosters respect and prevents tension from flaring into hostility. The disciplined application of force is more than a moral issue; it is an essential factor in operational success. The basis of the discipline that you and your Soldiers are expected to demonstrate when interacting with DCs is set forth in the mission's Rules of Engagement. Commanders rely on information operations, especially psychological operations, and related activities (public affairs and civil-military operations) to help favorably influence the local population's perceptions and attitudes.

US Force Responsibilities

Several individuals and units have major roles in US forces' interactions with DCs:

Theater Commander. In coordination with the State Department, the UN, US allies, and the host nation, the theater commander assigns DCs to the appropriate category, as listed earlier. Subordinate commanders—such as platoon leaders—must make sure not to inappropriately treat civilians in each category within the area of operations as EPWs.

Civil Affairs (CA). The major activity of CA is to plan and conduct DC operations. In so doing, their goal is to minimize civilian interference with military operations and to protect civilians from combat operations. CA personnel, in collaboration with military police and transportation officers, should coordinate with the international support community, host-nation military, and other allied military to mark safe routes for civilians to use. These marks must be made in languages and symbols that local civilians, US forces, and allied forces can all understand.

Military Police (MP). MP units are responsible for establishing routes, camps, and services for EPWs and civilian internees (CIs). CIs are individuals who are security risks or who need protection because they committed an offense against the detaining power (for example, insurgents, criminals, and other persons).

Psychological Operations. Psyops units are responsible for:

- Assessing the psychological impact of military operations
- Countering hostile propaganda
- Supporting information and awareness programs
- Supporting, planning, or conducting deception operations
- Providing target-audience intelligence and regional and language expertise
- Providing a means for distributing information and products that describe the intent of military operations
- Supporting the handling of EPWs and CIs.

After World War II, the Allies had to repatriate millions of dislocated civilians from dozens of countries. The dislocated civilians received passes from Allied military commanders printed in multiple languages to allow them to return home through military lines.

Noncombatant Evacuation Operations (NEOs)

You may sometimes be involved with civilians in a different way. If the environment poses a threat to civilians, it may be necessary to evacuate them to a secure area. Noncombatant evacuation operations (NEOs) relocate threatened civilian noncombatants from locations in a foreign nation to secure areas. Normally, these operations involve US citizens whose lives are in danger, either from the threat of hostilities or from a natural disaster. They may also include host-nation citizens and third-country nationals.

The threats may include impending natural disasters as well as combat operations. Army forces, normally as part of a joint task force, conduct NEOs to assist and support the State Department.

If they believe conditions warrant it, US ambassadors may launch an NEO in anticipation of a crisis. The environment may allow you to operate without threat, it may be uncertain, or it may be hostile. Direct military involvement in these evacuations is usually not required—the military usually plays a support role. The ambassador normally launches an NEO supported by the military when the local situation has deteriorated and the security of the evacuees is uncertain or when the environment is hostile. Such NEOs are usually conducted with little warning. Often American lives are in immediate danger.

Evacuation force commanders often have little influence over the local situation. They may not have the authority to use military measures to preempt hostile actions, yet must be prepared to protect the evacuees and defend the force. The imminent threat may come from hostile forces, general lawlessness, dangerous environmental conditions, or a combination of the three.

Evacuation operations differ from other military operations in that the US ambassador may direct the operation. Further, the order to evacuate is a diplomatic—rather than a military—decision, with extensive ramifications.

The Culture Maze

Each partner in a unified multinational action has a unique cultural identity. Military forces, civilian agencies, NGOs, and IOs approach operations from different perspectives. Factors such as national and organizational values, standards of social interaction, religious beliefs, and organizational discipline affect the partners' perspectives.

Partners with similar cultures and a common language face fewer obstacles to working together than do partners with differing backgrounds. Even such seemingly minor differences as dietary restrictions or relationships between officers and enlisted personnel can significantly affect military operations. As an Army officer, you will have to deal with cultural sensitivities and overcome diverse or conflicting religious, social, or traditional requirements.

Overcoming language barriers is among the greatest challenges. Unified action is often multilingual. Even when partners share a language, as US and British forces do, different terminology and jargon can hinder understanding.

In working with other US military services and civilian agencies, remember that they often have their own jargon and may not understand yours. Speak and write in plain English.

All participants must understand all spoken and written communication. This may require obtaining a translator's services. Translating orders adds time to planning, and errors in translation can cause mistakes or serious misunderstandings. Few translators available in the field have the combination of language, cultural expertise, and depth of doctrinal understanding necessary to do the job well.

Dedicated Army liaison and linguist teams can lessen, but not eliminate, this problem. To minimize, if not eliminate, cultural miscommunication, you must be clear and concise in speaking and writing. Clear, concise orders and briefings are easier to translate than complicated ones are. Ask your multinational subordinates for back briefs, just as you should do with your own Soldiers. This helps ensure that they understand your intent and their assigned tasks.

Critical Thinking

Why is it important to develop good relationships with NGOs, IOs, and other noncombatant elements in any operating environment? What actions can you take to do so?

CONCLUSION

As a platoon leader, you will deal with some or all of the nonmilitary organizations and groups described in this section. Misunderstandings between your platoon's Soldiers and one of these elements, even if unintentional, can have dire repercussions. It could lead to inflamed violence and international incidents that can embarrass the United States and even lead to the death of some of your Soldiers. It can make it harder for you to bring relief to suffering people and lead to unnecessary loss of civilian lives. Familiarity with the groups described in this chapter and willingness to work with them toward common goals will help you succeed in the full spectrum of military operations—whether they are combat, support, stability, or peace operations.

Key Words

international organizations (IOs)
nongovernmental organizations (NGOs)
civil-military operations center (CMOC)
psychological operations (Psyops)

Learning Assessment

1. Describe the role of nongovernmental organizations (NGOs).
2. Explain noncombatant evacuation operations (NEOs).
3. Identify and explain the major population concerns in military planning and operations.
4. Identify and describe the categories of displaced civilians (DCs).

References

Field Manual 1-02, *Operational Terms and Graphics.* 21 September 2004.

Field Manual 3-0, *Operations.* 27 February 2008.

Field Manual 3-05.30, *Psychological Operations.* 15 April 2005.

Field Manual 3.05-40, *Civil Affairs Operations.* 29 September 2006.

Field Manual 3-07, *Stability Operations.* 6 October 2008.

Field Manual 3-07.31, *Peace Operations: Multi-Service Tactics, Techniques, and Procedures for Conducting Peace Operations.* 26 October 2003.

Field Manual 27-10, *The Law of Land Warfare.* Change 1. 15 July 1976.

Field Manual 46-1, *Public Affairs Operations.* 30 May 1997.

Field Manual 100-8, *The Army in Multinational Operations.* 14 June 1997.

Field Manual 100-10-2, *Contracting Support on the Battlefield.* 4 August 1999.

Operation Provide Comfort. (n.d.). GlobalSecurity.org. Retrieved 1 December 2005 from http://www.globalsecurity.org/military/ops/provide_comfort.htm

Scott, S. (2008). Veterans Day a Time to Remember Civil War Soldiers. *Creators Syndicate, Inc.* Retrieved 5 November 2008 from http://www.creators.com/lifestylefeatures/travel-and-adventure/veterans-day-a-time-to-remember-civil-war-soldiers.html

Some Events Connected with the Life of Judith Carter Henry. (n.d.). Unpublished manuscript, Manassas National Battlefield Park, National Park Service. Retrieved 5 November 2008 from http://www.nps.gov/nr/twhp/wwwlps/lessons/12manassas/12facts3.htm

SUPPLY AND MAINTENANCE

Key Points

A slipping gear could let your M203 grenade launcher fire when you least expect it. That would make you quite unpopular in what's left of your unit.

Anonymous

Introduction

Even when everything else is going splendidly, **supply** and **maintenance** problems can drag your unit down. Your platoon might be the best trained unit in the brigade, but if you don't have the ammunition, food, and other supplies you need, or if your equipment doesn't work right, you won't get very far. Supply and maintenance— a key part of **logistics**—are what allow you to fight and win battles.

   This lesson will acquaint you with the basic supply and maintenance concepts and functions that you will need to know as a platoon leader carrying out your day-to-day assignments and missions.

supply

the acquiring, receiving, storing, and issuing of all classes of supply, except Class VIII (medical items), required to equip and sustain Army forces

Class VIII items are issued through medical channels.

maintenance

actions taken to keep materiel in serviceable, operational condition, and updating and upgrading its capability— including performing preventive maintenance checks and services; recovering and evacuating disabled equipment; diagnosing equipment faults; substituting parts, components, and assemblies; exchanging serviceable materiel for unserviceable materiel; and repairing equipment

logistics

the science of planning and carrying out the movement and maintenance of forces

Supply

Supply is the process of providing all items necessary to equip, maintain, and operate a military unit. Supply shortages (especially of ammunition, fuel, and repair parts) can prevent your unit from accomplishing its mission. As a platoon leader, you must know all about the resources your platoon needs—the type, quantity on hand, location, condition, and availability. You must also know the current rate of use and be able to estimate future consumption rates based on the situation you face.

Successful Army operations exhibit five essential characteristics or tenets: initiative, agility, depth, versatility, and synchronization. These characteristics are as important to supply operations as to any other. Table 6.1 explains how these characteristics relate to supply operations.

TABLE 6.1	Tenets of Army Operations and Supply	
Tenet	**Definition**	**Supply Applicability**
Initiative	Setting or changing the terms of battle by action.	Thinking ahead and anticipating future requirements while planning supply needs beyond the current operation.
Agility	The ability of friendly forces to act faster than the enemy.	Physical agility depends upon the right quantity of supplies, both enough but not too much. Mental agility can be affected by low morale or poor health, which can be caused by the wrong amount of supplies, for example: food, water, clothing.
Depth	The extension of operations in space, time, and resources.	Proper use of supplies plays a critical role in achieving and maintaining momentum in the attack and elasticity in the defense.
Versatility	The ability to tailor forces and move rapidly and efficiently from one mission to another.	The successfulness of moving from one mission to another will not be efficient if the supplies are not in the right place at the right time.
Synchronization	The arrangement of battlefield activities to produce maximum combat power at the decisive point.	If supply support, especially ammunition and fuel, is not correctly synchronized, units will fail to achieve maximum combat power at critical moments.

The Character of Logistics

A successful logistics operation also has several defining characteristics. These characteristics seldom have equal influence on an operation, but identifying them during the planning process will provide you with a guide for analytical thinking and prudent planning.

Anticipation. Foresee future operations and identify, accumulate, and maintain the assets and abilities to support those operations.

Simplicity. Avoid complexity in both planning and executing logistical functions. Mission orders, drills, and standing operating procedures (SOPs) contribute to simplicity.

Responsiveness. Get the right supplies and other support functions to the right place at the right time.

Economy. Provide appropriate support without excess. You need to judge economy in prioritizing and allocating resources.

Flexibility. Make sure organizational structure and logistical procedures are adaptable to changing situations, missions, and operations.

Integration. Coordinate logistical operations with the other missions and components of the organization.

Attainability. Make sure you have at least the minimum quantity of supplies and services required to begin an operation.

Sustainability. Ensure that you can maintain continuous support to all phases of operations.

Survivability. Shield logistics functions from destruction.

Improvisation. Ensure that you can make, fabricate, arrange, or invent what is needed from available supplies. This is not, however, an acceptable alternative to proper planning.

The Service Support Functions

The six service support functions are *man, move, maintain (fix), arm, fuel,* and *sustain.* If your unit is to accomplish its mission, these functions must occur. It's the job of logisticians and staff officers to ensure that the six supply functions are adequately addressed in paragraph 4 of an Operation Order. Reviewing these functions gives you, the platoon leader, a way to make sure your platoon has what it needs to get the job done. At the platoon level, you and your platoon sergeant are responsible for addressing these needs.

TABLE 6.2	Classes of Supply	
SUPPLY		**CLASS**
Subsistence		I
Clothing, individual equipment, tentage, organizational tool sets and tool kits, hand tools, maps, and administrative and housekeeping supplies and equipment		II
POL (package and bulk): petroleum fuels; lubricants; hydraulic and insulating oils; preservatives; liquid and compressed gases; bulk chemical products; coolants; deicing and antifreeze compounds, together with components and additives of such products; and coal		III
Construction materials, including installed equipment and all fortification and barrier materials		IV
Ammunition of all types, chemical and special weapons, bombs, explosives, mines, fuses, detonators, pyrotechnics, missiles, rockets, propellants, and other associated items		V
Personal demand items (nonmilitary sales items)		VI
Major end items: a final combination of end products that are ready for their intended use, for example, tanks, launchers, mobile machine shops, and vehicles		VII
Medical material, including medical-peculiar repair parts		VIII
Repair parts (less medical-peculiar repair parts): all repair parts and components, to include kits, assemblies, and subassemblies—repairable and nonrepairable—required for maintenance support of all equipment		IX
Material to support nonmilitary programs, such as agricultural economic development, not included in classes I through IX		X

Classes of Supply

The Army divides supplies into 10 categories or *classes,* which you should memorize. Table 6.2 describes and defines them.

You will be expected to address these supply categories in your unit on a day-to-day basis, particularly when you are in the field. When your platoon needs something, you'll use this terminology to notify your commander and supply channels.

Critical Thinking

Do you think that there is one class of supply that is more important than another in supporting Soldiers in the Contemporary Operating Environment (COE)? Why? Do you think there is a less important class of supply in the COE? Why?

Accounting for Supplies and Equipment

When you assume leadership of your platoon, you also assume responsibility for the Army property assigned to it. The two categories of Army property are *real property* and *personal property*. Real property includes land and structures. Personal property includes capital equipment and other nonexpendable supplies.

All Army property, except real property, is classified for accounting purposes as *nonexpendable, durable,* or *expendable.*

Nonexpendable property is property that is not consumed when used and that keeps its original identity after use. Nonexpendable items must be accounted for in the unit **property book** after the higher issuing unit has issued them. These items include all nonconsumable major end items.

Durable property is personal property that is not consumed in use and that keeps its original identity. Durable items need not be accounted for in the property book after issue from the stock record account. They do, however, require control using hand receipts when issued to the user. This includes hand tools with a unit price greater than $50 and cellular phones, pagers, and BlackBerry units with a unit cost of less than $1,000.

Expendable property is personal property that is consumed in use or that loses its identity in use; or property with a unit price less than $100 that is neither consumed in use nor otherwise classified as durable or expendable. Expendable items require no formal accountability after issue from a stock record account. They include oil, paint, fuel, repair parts, audiovisual products, training devices, and training aids.

property book

a formal set of property accounting records and files maintained at the user level—it is used to record and account for all nonexpendable and other specially designated property issued to that activity

To ensure good fiscal discipline and responsible use of taxpayer dollars, you must closely account for expendable items that are popular or can easily be pilfered (such as USB drives, computer speakers, dry-erase boards, terrain model kits, laser pointers, and so on). You should use hand receipts for issuing these expendables to Soldiers to ensure accountability, so the unit does not continue to repurchase these items over time.

Accounting Tools

The Army is well known for accounting for its property—few if any organizations have so much to keep track of. If your platoon is picking up food, ammunition, and equipment, how do you keep track of it? As you may deduce from the information above, there are various methods of accounting for the different categories of supply and property, and various forms for doing so.

FM 10-27-4, *Organizational Supply for Unit Leaders,* says: "Property responsibility is the obligation of a person to ensure that government property entrusted to his possession, command, or supervision is properly used and cared for and proper custody and safekeeping are provided." To ensure this happens, the Army has various inspection and auditing procedures.

Your company commander will sign for all property assigned to the company, including all platoon property. Your commander will then pass platoon property down to you and the other platoon leaders using subhand receipts. As a platoon leader, your job is to make certain that you "subhand receipt" the platoon's property down to your platoon sergeant and squad leaders and ensure that your noncommissioned officers (NCOs) and Soldiers keep adequate records of the property under your control. You do this using *property books; hand receipts; inventory lists; transfer documents;* and operational, prescribed, and basic *load lists.*

HAND RECEIPT/ANNEX NUMBER For use of this form, see DA PAM 710-2-1. The proponent agency is DCS, G-4.		FROM:		TO:						HAND RECEIPT NUMBER				
FOR ANNEX/CR ONLY	END ITEM STOCK NUMBER	END ITEM DESCRIPTION		PUBLICATION NUMBER					PUBLICATION DATE	QUANTITY				
STOCK NUMBER a.		ITEM DESCRIPTION b.			* c.	SEC d.	UI e.	QTY AUTH f.	g.	QUANTITY				
										A	B	C	D	E	F

* WHEN USED AS A:

  HAND RECEIPT, enter Hand Receipt Annex Number
  HAND RECEIPT FOR QUARTERS FURNITURE, enter Condition Codes
  HAND RECEIPT ANNEX/COMPONENTS RECEIPT, enter Accounting Requirements Code (ARC).

PAGE _____ OF _____ PAGES

Figure 6.1 DA Form 2062

Hand Receipts

A *hand receipt* is a listing of nonexpendable and durable items (other than components) that have been issued to an individual, section, or unit. The signature on a hand receipt establishes direct responsibility for that item. Hand receipts are also accountable records of all nonexpendable and durable property. Manual systems use the DA Form 2062 as hand receipt documents to account for property at company, unit, or activity level. The form assigns responsibility to the supervisor and user. Instructions for preparing the DA Form 2062 (Figure 6.1) are found in DA PAM 710-2-1. Automated systems use machine listings as hand receipt documents. These are prepared and maintained according to DA PAM 710-2-1 and the system-end user's manual.

Inventory Lists

The property book officer (PBO) or responsible officer may encounter a situation where it is impractical to assign further responsibility for property. For example, this could happen in the case of multiple-use classrooms or day rooms used by more than one unit. In this case, you may manage the property using an automated list or DA Form 2062 as an inventory list. When using the inventory list method, the PBO or responsible officer must inventory the property semiannually. You prepare and manage the list according to the provisions of DA PAM 710-2-1.

Issue, Turn-In, and Transfer Documents

A PBO or responsible officer may use DA Form 3161 in many different situations. DA PAM 710-2-1 provides specifics for preparation of the form. You can use DA Form 3161 (Figure 6.2) as an issue, turn-in, or transfer document.

REQUEST FOR ISSUE OR TURN-IN
(DA PAM 710-2-1)

ISSUE / TURN-IN	SHEET NO. 1	NO. SHEETS	1. REQUEST NO.	2. VOUCHER NO.
3. SEND TO:	4. DATE MATERIAL REQUIRED *(YYYYMMDD)*	5. DODAAC	6. PRIORITY	7. ACCOUNTING/FUNDING DATA
8. REQUEST FROM:	9. END ITEM IDENT	9a. NAME/MANUFACTURER	9b. MODEL	9c. SERIAL NO.
* CODE — ISSUE: *I-Initial*, *R-Replacement*; TURN-IN: *FWT-Fair Wear And Tear*, *RS-Report of Survey*, *EX-Excess*, *SC-Stmt of Charges*, *LT-Lateral Transfer*		10. PUBLICATION		11. JOB ORDER NO.

12. ITEM NO. *a*	STOCK NO. *b*	ITEM DESCRIPTION. *c*	UNIT OF ISSUE *d*	QUANTITY *e*	CODE* *f*	SUPPLY ACTION *g*	UNIT PRICE *h*	TOTAL COST *i*	*j.* POSTED DATE *(YYYYMMDD)*	BY
							SHEET TOTAL		GRAND TOTAL	

13. ISSUE/TURN-IN "QUANTITY" COLM IS REQUESTED	DATE *(YYYYMMDD)*	BY	14. ISSUE QTY IN "SUPPLY ACTION" COLUMN	DATE *(YYYYMMDD)*	BY	15. REC QTY IN "SUPPLY ACTION" COLUMN	DATE *(YYYYMMDD)*	BY

Figure 6.2 DA Form 3161

Request for issue and turn-in document. Units not using an automated system may use DA Forms 3161 and 3161-1 (continuation sheet) to request supplies. You may also use these forms to document turning in of items to the PBO or other activity. They are valid for only 30 days, however, at which time the hand receipt must be updated.

Note:

It is important for you to establish a monthly routine during which you and your platoon sergeant review DA 3161s to ensure that a Soldier does not depart the unit via permanent change of station (PCS) or expiration term of service (ETS) without first turning in or clearing his or her outstanding DA 3161. If a Soldier departs with an outstanding DA 3161 and you cannot account for the items on the DA 3161, you must immediately inform your commander so he or she can initiate a Report of Survey investigation in order to determine who is responsible for the missing property. Failing to keep track of your DA 3161s can result in you having to pay for the missing property.

Transfer document. Units can transfer items laterally to other units when authorized or directed by the appropriate level of command, depending on the type of property involved. These transfers can be posted to the hand receipt using the DA Form 3161. Procedures for lateral transfer actions are contained in DA PAM 710-2-1.

Note:

It is prudent to always keep previous copies of your unit property book and receipts, especially when they have been updated. This gives you a track record of property that was removed or transferred from your property book. Transfer documents can become lost, and transfers can be posted incorrectly. This can result in property that your predecessor turned in or dropped still showing up on your automated property books—sometimes years after the platoon leader who preceded you turned in the property.

Hand Receipt Annexes

Hand receipt annexes are used between the PBO and primary hand receipt holders and between primary hand receipt holders and sub-hand receipt holders. When an item with components is issued on a hand receipt or sub-hand receipt, you must record any shortage of nonexpendable or durable components on a hand receipt annex. You prepare the hand receipt annex, also known as a shortage annex, in two copies. You show the shortage of any component of a major end item, set, kit, or outfit on a hand receipt annex for that item of property. The PBO and primary hand receipt holders record shortages of nonexpendable components. The person who maintains the document register for durable items (the PBO or company commander) will record durable items on the hand receipt annex.

Loads

Loads are the amount of durable and expendable supplies units keep to sustain their operations. You must track these for your unit to maintain a wartime footing. You do not keep loads of Class VI, VII, and X. There are generally three types of loads:

Basic loads. Basic loads are Major Army Command (MACOM)-designated quantities of Class I through V and VIII supplies. These allow a unit to conduct its combat operations. During peacetime, you use items from the basic load only when no operational loads are available. You must be able to move your basic load into combat using your unit's regular assigned transportation in a single lift. You must maintain or replace those basic load items that can deteriorate or that have a limited shelf life.

Operational loads. Operational loads are quantities of Class I through V and VII supplies that the unit keeps to sustain its peacetime operations for a given time. Operational load quantities are based on the history of your unit's supply use. You may move these supplies into combat if transportation is available after essential lift requirements have been met.

Prescribed loads. Prescribed loads are quantities of Class II, IV, VIII, and IX repair parts you keep to support the unit's maintenance program. The quantities are based on items designated by the command and the unit's demand history. You may move these parts into combat if transportation is available after essential lift requirements are met. AR 710-2 gives Army policy and sets the numbers of lines and quantities of prescribed loads a unit is authorized.

Financial and Property Responsibility

By law, a person responsible for a loss must reimburse the US government for government property that was lost, damaged, or destroyed because of negligence or willful misconduct. A person may be held liable by his or her admission or as the result of an investigation. The concept of financial responsibility is that the person responsible for the loss should make reparations for the loss rather than be punished for it. Soldiers can be charged the full amount of the loss when personal arms or equipment are lost.

Commanders who maintain separate property books at the company level are *accountable officers*. Accountable officers are liable for the full amount of the loss unless they can prove they were not at fault.

Every Soldier has an obligation to ensure that government property entrusted to his or her possession, command, or supervision is properly used and cared for, and to provide proper custody and safekeeping. AR 710-2 requires that someone be assigned direct responsibility for each nonexpendable and durable item on hand in the unit. There are three types of responsibility based on position within the organization:

Command responsibility is a commander's obligation to ensure the proper care, custody, and safekeeping of all government property within the command. You have this command responsibility for your unit property whether you have signed for it or not. You must personally ensure the security of all unit property, whether it is in storage or in use. For example, you must provide a secure place for mechanics to store the tool kits issued to them. If you have not done so, and an item is lost, you could be held liable for the loss. You must also ensure proper supervision to make sure the mechanic is using the tool kit correctly.

Supervisory responsibility is a supervisor's obligation to ensure the proper use, care, and safekeeping of government property issued to or used by subordinates. Supervisors can be held liable for losses their subordinates incur.

Personal responsibility is your Soldiers' responsibility for all weapons, hand tools, and organizational clothing and individual equipment issued to them for their use. They are responsible whether they signed for the property or not. For example, when the tool kit is issued, the mechanic assumes personal responsibility for it and all items in it. The mechanic must take proper care of the kit and secure it in the assigned storage area when he or she is not using it. If the mechanic forgets to secure the kit and it is lost, he or she is responsible for the loss.

Critical Thinking

In what light do you think your rater, intermediate rater, senior rater, and other officers will regard you if you are found financially liable for property that you or your Soldiers failed to maintain accountability for?

Command Maintenance Program

The Army is transforming and reorganizing for 21st-century operations. As a new officer, you are expected to give day-to-day guidance to Soldiers. One of the most important areas for which you are responsible is maintenance. Specifically, you are responsible for ensuring that your Soldiers are performing the required **preventive maintenance checks and services (PMCS)**.

preventive maintenance checks and services (PMCS)

the checks, service, and maintenance performed before, during, and after any type of movement, or before the use of all types of military equipment

Critical Thinking

Do you think that PMCS is a necessity for a commander's maintenance success? Why?

A **command maintenance program** ensures that all vehicles and equipment receive thorough weekly inspections. Performing preventive maintenance checks and services (PMCS) to standard is the key to identifying and repairing faults—and to reducing the risk of equipment damage or personal injury due to failures. There is much more to the maintenance program than meets the eye, however. All the parts work in unison to achieve the final objective.

Soldiers do best the work that their commanders and supervisors check and recognize. As much as possible, you should use command review and recognize high achievement to make your organization's maintenance program successful and to reward deserving Soldiers.

command maintenance program

a program in which all vehicles and equipment receive thorough weekly inspections

Individual/Soldier recognition—Effective leaders use many methods to recognize individual Soldier achievement and success. One method, for example, is through DA PAM 672-5, which specifies that commanders (lieutenant colonel or higher) can award driver and mechanic badges, with appropriate bar(s), to persons who demonstrate a high degree of ability in equipment operation or mechanical maintenance.

Unit Recognition—The Department of the Army has established a unit recognition program—The Chief of Staff Army Award for Maintenance Excellence Program (AAME)—that can serve as the cornerstone of major subordinate command (MSC) unit maintenance recognition programs.

The objectives of the AAME program are to:

- improve and sustain maintenance readiness
- assess the status of total unit maintenance operations
- improve efficiency and reduce waste
- recognize outstanding accomplishments and initiatives, ensure that the best units compete, and promote competition at MSCs, Department of the Army HQ (HQDA), and Defense Department (DoD) levels. Commanders administer this program within the guidelines in AR 750-1 (Chapter 8 and appendix D). The four component competition areas are:
  - Active Army table of organization and equipment (TOE)/MTOE unit
  - Army National Guard TOE/MTOE unit
  - US Army Reserve TOE/MTOE unit
  - Table of distribution and allowances (TDA) unit (any component).

Representatives from winning units receive their awards and recognition at a ceremony conducted annually by Chief of Staff Army in Washington, DC, usually during June or July. A winning unit and runner-up are selected for each of the three categories (light, medium, heavy) for each of the four components listed above. Each of the 12 winning units can select up to three Soldiers to represent their unit at the awards ceremony.

The Secretary of Defense Maintenance Award Program—This program annually recognizes the top six maintenance units across all services. An HQDA board selects Army nominees from among units that competed and were selected as AAME winners. The top AAME winners are the Army's nominees for the DoD Maintenance Award. An Army unit must compete in the AAME to be nominated to the DoD Maintenance Award Program. One of the six units, from all services, is then selected as the best overall throughout DoD and is awarded the Secretary of Defense Maintenance Award.

Unit Safety Management and Maintenance Advisory Messages—Maintenance-related accidents are responsible for 20 percent of all the military's on-duty injuries. Accidents reduce a unit's effectiveness, negatively affect morale and discipline, and degrade operational abilities. Goals of all unit safety programs include increased safety awareness and preventing accidents. Army policy for the prompt notification of field commands regarding safety issues is found in AR 750-6.

Unit Level Safety Inspections—These inspections, including "management by walking around," are a must for a unit's effective maintenance safety program. A dirty or disorderly shop should be a supervisor's first indicator of unsafe maintenance operations. The following questions can assist you in getting your unit started on the right track:

- Does the supervisor have a written, formal accident-prevention plan that is compatible with the mission and the function of the organization? Are unit personnel aware of and actively implementing the plan?
- Does the unit have a current, complete, and clearly defined safety standing operating procedure based on AR 385-10, AR 385-40, AR 385-55, and DA PAM 385-1?
- Does the unit conduct safety meetings regularly?
- Is the unit leader/activity supervisor directly involved in the unit accident prevention and safety awareness program?
- Is there a safety officer designated on orders? Are duties specified? Are duties actually accomplished or just given lip service?

Ground Safety Notification System—This is the Army's system used to distribute high-, medium, and low-category safety messages to the field. When safety conditions surrounding operation or custody of Army equipment reach risk levels, or accidents occur, the Army Materiel Command (AMC) or other Army-level organizations may send urgent messages to field users to alert them to potential hazards. Army policy for these notifications is contained in AR 750-6.

Test, Measurement, and Diagnostic Equipment—**Test, measurement, and diagnostic equipment** (TMDE) is any system or device that can evaluate the operational condition of equipment or subsystems and their potential malfunctions, or that can determine whether a part or item is installed within specifications. Using TMDE, the Soldier maintainer can identify and/or isolate actual or potential malfunctions. The regulation covering TMDE is AR 750-43. It explains the Army TMDE Calibration and Repair Support Program and requires units to appoint, on orders, a TMDE calibration coordinator.

test, measurement, and diagnostic equipment (TMDE)

any system or device that can evaluate the operational condition of equipment or subsystems and their potential malfunctions, or that can determine whether a part or item is installed within specifications

Commanders must ensure that a unit TMDE calibration coordinator is on orders at all times and is actively supporting commands as outlined in AR 750-43.

Technical Publications—The primary source of publications is the US Army Publishing Agency (www.apd.army.mil). Units and activities can use, review, print, or download the electronic versions of publications and forms at this site or be linked to another library.

Tools and Tools Improvement Program—You can't perform maintenance correctly without proper tools. This includes not only tools authorized by the modified table of organization and equipment (MTOE), but also special tools authorized by repair parts and equipment technical manuals (TMs).

Maintenance Assistance & Instruction Team—The primary purpose of the **Maintenance Assistance and Instruction Team (MAIT)** is to upgrade Army materiel and units to a high state of readiness. It does this by providing effective and responsive assistance and instruction to units and activities. MAIT operates as a decentralized program.

maintenance assistance and instruction team (MAIT)

a team that upgrades Army materiel and units to a high state of readiness

Army Oil Analysis Program—The purpose of the Army Oil Analysis Program (AOAP) is to assist you by providing an oil analysis and report service on combat equipment, aircraft, and watercraft. AOAP analysis can detect potential equipment component failures, enabling you to prevent catastrophic failure of equipment if you take prompt management action. You can do this by applying an oil-change policy. AOAP analysis identifies lubricant conditions by evaluating equipment oil samples. A well-run AOAP can save oil, repair parts, labor, and organization funds. You should use proper sampling procedures and submit samples promptly.

Facilities, Shop Layout, and Production Enablers—Shop organization and layout are key factors in the efficient functioning of a unit's maintenance operation, both in the field and in garrison. Inadequate facilities or inefficient layout can reduce mission output and, operationally, ready rates for the unit. They waste Soldier maintainers' time and effort.

Warranty Programs—The overall policies and procedures for the Army Warranty Program for IT are contained in AR 700-139. A leader of Army maintainers should be aware of the following:

- Major subordinate commands (MSCs) acquire warranties only when they are in the Army's best interest. Acquiring commands or activities are to establish local warranty implementation procedures.
- AR 700-139 requires each MSC to appoint a warranty control officer (WARCO).
- In warranty applications, unit readiness and mission effectiveness take priority over warranty actions. If the Maintenance Allocation Chart (MAC)-Code F maintenance provider is not able to get action under a warranty in a timely manner, the maintenance provider should repair the equipment first and initiate settlement action later in accordance with local SOPs and AR 700-139. Make sure to notify the supporting WARCO immediately when equipment must be fixed first and the warranty settled later.

Maintenance Regeneration Enablers—Maintenance regeneration enablers are separate stocks of end items, held by accountable officers at different logistics levels, for timely issue of assets to units and organizations. You use maintenance regeneration enablers when internal maintenance actions cannot meet readiness timelines.

Unique Item Tracking Program—The **unique item tracking (UIT)** program requires the visibility and tracking by serial number of selected items and installed components as outlined in DOD 4140.1-R, DOD 4000.25-2-M, and AR 710-3. The objective of the UIT program is to maintain the visibility of each uniquely identified asset for inventory control and/or engineering analysis. Security, accountability, safety, maintenance, operational readiness, warranty applicability, and other areas that may benefit from the tracking process will be subsets of the inventory control or engineering-analysis functions.

unique item tracking (UIT)

a program that requires the visibility and tracking by serial number of selected items and installed components

Logistics Assistance Program—The Commanding General, AMC, manages the worldwide Logistics Assistance Program (LAP). Each AMC MSC provides technical and logistical assistance to the field level of maintenance for the equipment for which they are responsible. This logistics and technical assistance is provided through a number of logistics assistance officers (LAOs) and logistics support elements (LSEs) that are strategically located in all major Army geographic areas.

LAOs are typically found on Army installations in established areas and can perform such assistance services such as:

- tracking down the exact status of a critical requisition
- finding a critical part, module, or subassembly
- helping resolve systemic supply and maintenance problems
- providing assistance on warranty issues
- coordinating special training on new equipment
- providing onsite technical and logistics training when needed
- spot checking total package fielding for the MSC.

KNOW THE
**LEADERSHIP
PRINCIPLES**
FOR MAINTENANCE

PMCS Responsibility of Key Unit Personnel

To make sure your Soldiers are performing the required PMCS, you must first understand the importance and key role of PMCS, and the responsibility of key unit personnel.

Preventive maintenance checks and services are the starting points for Army maintenance operations and the condition-based maintenance-plus (CBM+) approach to maintenance. Operators and crews must observe equipment performance and condition, then document and report what they detect. The equipment technical manuals require conducting PMCS before, during, and after operations. This is the foundation for the Army maintenance program. Higher-level maintenance operations ensue from unit use of equipment and PMCS.

Operators and crews observe and document equipment and weapon-system performance and condition against established standards. They report problems that degrade the items before conditions become catastrophic. A key point to remember here is that you must provide adequate time in training and operations schedules for Soldiers to perform PMCS. You must then allow the time needed to diagnose and correct equipment faults and to forecast the items' future serviceability. This is the essence of CBM+.

PMCS is the essence of CBM+.

Soldiers in Army field organizations who preserve the operational condition and inherent reliability of equipment are the most critical building block in the Army maintenance system. Your maintenance teams will succeed when they sustain equipment with operational ready rates at required levels while achieving the Army Maintenance Standard for assigned and attached equipment.

Your leadership and commitment strengthen the probability of success of any task, mission, or course of action. Maintenance tasks require your effective leadership to get the job done in accordance with policy and the best way possible.

Commanders of organizations at battalion level and below—and commissioned, warrant, and noncommissioned officers within those organizations—occupy the most critical positions in the Army maintenance process. If you give maintenance operations proper priority in regard to overall unit mission requirements, you will improve the chances for your unit's success and mission accomplishment. Other supportive behaviors and actions are also required of you if the maintenance mission is to succeed.

You must implement the policies contained in AR 750-1; the procedures contained in DA PAM 750-8 and DA PAM 738-751; the automated processes contained in the Unit Level Logistics System-Ground (ULLS-G), ULLS-Aviation (A), and Standard Army Maintenance System-Enhanced (SAMS-E); and in succeeding generations of maintenance software. Each level of command has its assigned and implied responsibilities. You need dedication, teamwork, and coordination to accomplish the maintenance mission and implement it correctly.

You provide the leadership link to the operators and crews and support the achievement of the Army Maintenance Standard by:

- preparing for and ensuring that your subordinates fully participate in scheduled preventive-maintenance periods
- attending, leading, and supervising preventive-maintenance operations
- being technically competent
- checking and updating SOPs
- knowing your responsibilities for your areas of supervision and maintenance-operations procedures
- enforcing the Army Maintenance Standard for the equipment you are responsible for, and ensuring that subordinate supervisors, leaders, crews, and operators/users have the desired sense of "ownership"
- training operators and crews to operate equipment and perform PMCS properly
- enforcing safety rules
- recording and reporting maintenance data in accordance with DA PAM 750-8 and DA PAM 738-751
- informing your chain of command when there is not sufficient time, personnel, funding, tools, TMs, or other maintenance means to maintain equipment as required.

Battalion Maintenance Officers

Not all battalions have a *battalion maintenance officer (BMO)*. For those battalions that do, they are an important player in the maintenance arena and an asset to the commander.
A BMO:

- controls the total maintenance effort of the maintenance platoon
- establishes the maintenance priorities to support the commander's mission
- gives the commander accurate equipment status reports
- evaluates the unit's PMCS operation
- enforces the commander's maintenance standards
- assists the commander in planning tactical maintenance support
- fully understands status reporting for materiel and unit equipment
- works closely with your support in maintenance and supply activities
- continually monitors operations, recommends changes, and improves maintenance
- ensures that sufficient copies of TMs and lubrication orders (LOs) are available for Soldiers to perform PMCS and organizational maintenance on the unit's organic equipment.

The purpose of Army maintenance operations is to generate and regenerate combat power and to preserve the capital investment in combat systems and equipment over time to enable training and support all assigned missions.

DA Form 5988-E/DA Form 2404

DA Form 5988-E is a worksheet on which Soldiers record equipment condition and maintenance. The data on the form are divided into three sections:

Section 1: Equipment Data—The top section of the form gives basic information from the ULLS-G equipment data file and maintenance master data file. The main purpose of this data is to identify the vehicle.

Section 2: Parts Requested—This section identifies all parts on order for the vehicle. The document control register in ULLS-G reflects this information also.

Section 3: Maintenance Faults—This section shows a list of vehicle faults from the ULLS-G maintenance fault file. This includes all uncorrected faults for the equipment. This is where the driver also enters any new faults found during PMCS inspection.

The date in the upper left corner is the date the form was printed. Your unit SOP will give an interval for updating the form. If your unit SOP requires updating the form weekly, the date must be no more than a week old. The vehicle crew must know the current status of requested repair parts and previously reported faults. *The most important thing DA Form 5988-E does is to show equipment status.* The form and information must be current. An outdated form may not give the current vehicle status. Besides making routine updates, update the form immediately when vehicle mission-capable status changes.

The battalion equipment numbering system should be part of your unit SOP. For example, if someone told you that D2S is not mission capable, how would you know which part of the system is affected? Per the SOP, the "S" means that it is a weapon system; however, if the SOP states that all M240 machine guns will use the administration number suffix "S," then you would also know the weapon subsystem model. The "S" is an example only—your SOP may use other suffixes.

You keep DA Form 5988-E in the equipment record folder. There must be a form for each reportable item of equipment in the system. The crew must perform PMCS for each subsystem supporting the equipment's role to shoot, move, and communicate. In most cases, there will be more than one form in the record folder, since the crew must prepare a separate form on each subsystem. When the PMCS is done, Soldiers record it on the DA Form 5988-E with the same administration number as the vehicle. The vehicle administration number is the bumper number.

The lower two sections of the form list repair parts and equipment faults. When an operator discovers a fault that is not correctable at operator level, he or she writes this fact on the form just below the last entry. The next several examples will walk you through finding faults, recording them on the DA Form 5988-E, and updating the form when the faults are corrected.

DATE: 07-AUG 96	EQUIPMENT MAINTENANCE AND	DA FORM 5988-E
	INSPECTION WORKSHEET	
W45U7D	D TROOP, 1/10 ARMOR	

-- EQUIPMENT DATA --

ADMIN NUM: D2   EQUIP SERIAL NUM: MSJ21639MBB
EQUIP MODEL: M113A2   REGISTRATION NUM: MC03MW
EQUIP NOUN: CARRIER PERSONNEL   TYPE INSPECTION: DAILY
EQUIP NSN: 2340010684077

	NUMBER	DATE	CHANGE NUMBER
PUBLICATION:	TM 9-2350-261-PMC	04/87	00
PUBLICATION:	TM 9-2350-261-10	07/90	00

INSPECTORS LIC #: _____   TIME: _____   SIGNATURE: _____   TIME: _____

-- PARTS REQUESTED --

FAULT	DOC NUM	NON	NOUN	QTY DUE/ REC	STATUS DATE	DATE COMP	PRI	DLC

-- MAINTENANCE FAULTS --

ITEM NUM	FAULT DATE	FAULT STATUS	FAULT DESCRIPTION	CORRECTIVE ACTION	HRS	OPER LIC #
			7-AUG-96			F5294

Figure 6.3a DA Form 5988-E

Look at this DA Form 5988-E carefully. It represents a perfect vehicle. There are no faults noted, so no repair parts are required. You will probably never see a form like this again. Assume that during the before-operations PMCS on 7 August 1996 the operator finds no new faults. He or she enters the date the before-operations PMCS was completed in the FAULT DESCRIPTION column to indicate that a before-operations PMCS was conducted and no faults were found.

After the before-operations PMCS are complete, the operator makes the during-operations PMCS while the vehicle is operated and at halts. At the conclusion of the day's operation, the driver must perform the after-operations PMCS. After completing the during- and after-operations PMCS, the operator enters his or her driver's license number in the OPER LIC# column. This DA Form 5988-E for vehicle D2 indicates that the after-operations PMCS are complete and no new faults were found (Figure 6.3a).

There must be a new entry for each day the vehicle is operated and the operator reports no new fault. Soldiers can use the same form until it is updated or a new fault is reported.

On 8 August 1996, the operator of vehicle D2 discovers new faults and lists them on DA Form 5988-E as indicated on the adjacent form (Figure 6.3b). The driver may enter the new faults at the bottom of the form if the equipment is not going to be dispatched. Otherwise, the new faults must be entered on DA Form 2404 or a DA Form 5988-E continuation sheet. The operator discovered the faults during the before-operations PMCS because there is no date in the fault description column to indicate that the before-operations

```
 ┌──────────────────────────────────────────────────────────────────────────┐
 │ DATE: 07-AUG 96        EQUIPMENT MAINTENANCE AND        DA FORM 5988-E      │
 │                          INSPECTION WORKSHEET                               │
 │        W45U7D            D TROOP, 1/10 ARMOR                                │
 │ ---------------------------------- EQUIPMENT DATA ------------------------- │
 │   ADMIN NUM: D2                          EQUIP SERIAL NUM: MSJ21639MBB       │
 │   EQUIP MODEL: M113A2                     REGISTRATION NUM: MC03MW           │
 │    EQUIP NOUN: CARRIER PERSONNEL          TYPE INSPECTION: DAILY             │
 │    EQUIP NSN: 2340010684077                                                 │
 │                   NUMBER          DATE              CHANGE NUMBER           │
 │  PUBLICATION: TM 9-2350-261-PMC   04/87                 00                  │
 │  PUBLICATION: TM 9-2350-261-10    07/90                 00                  │
 │ INSPECTORS LIC #: F5294    TIME: _____  SIGNATURE: _____  TIME: _____    │
 │ ---------------------------------- PARTS REQUESTED ----------------------   │
 │                                                                            │
 │ FAULT  DOC NUM  NON    NOUN    QTY      STATUS     DATE                      │
 │                                DUE/ REC  DATE      COMP      PRI     DLC     │
 │ ---------------------------------- MAINTENANCE FAULTS -------------------    │
 │ ITEM    FAULT    FAULT    FAULT          CORRECTIVE            OPER          │
 │ NUM     DATE     STATUS   DESCRIPTION    ACTION        HRS     LIC #         │
 │                                7-AUG-96                        F5294         │
 │   2    8-AUG 96    X     ENGINE TEMP GA INOP                                 │
 │   8    8-AUG 96    /     HEADLIGHT INOP                                      │
 │  10    8-AUG 96    /     SEAT CUSHION TORN                                   │
 └──────────────────────────────────────────────────────────────────────────┘
```

Figure 6.3b DA Form 5988-E, continued

PMCS were completed on 8 August 1996. After entering the new faults on the form, the operator's driver's license number is entered in the INSPECTORS LIC# block.

The ITEM NUM is the PMCS check number from the TM-10. The numbers will be in ascending order if the driver performed the PMCS in TM sequence.

DA Form 2404

DA Form 2404 (Figure 6.4) is a manual form you use to report new equipment faults. The main difference between this form and DA Form 5988-E is that the operator must complete the form heading on the DA Form 2404; otherwise, the form is the same. The operator enters the date in the lower part of the form when the before-operations PMCS are completed and no new faults are found. The operator also initials the right column of the form when no new faults are discovered in during- and after-operations PMCS. When he or she discovers new faults, the operator enters usage in block 4 and the date in block 5 at the top of the form and signs the form. The operator must give the signed and dated form to his or her immediate supervisor and start a new form.

The ULLS-G computer produces the DA Form 5988-E continuation sheet. Soldiers may use it to report new faults when the vehicle is dispatched. When the ULLS-G clerk updates DA Form 5988-E, the system will print another DA Form 5988-E continuation sheet if requested. The advantages of this form over the DA Form 2404 is that the driver does not have to prepare the form heading, and the form is always available as long as the unit has a printer.

EQUIPMENT INSPECTION AND MAINTENANCE WORKSHEET

For use of this form, see DA PAM 738-750 and 738-751; the proponent agency is DCSLOG

1. ORGANIZATION					2. NOMENCLATURE AND MODEL		

3. REGISTRATION/SERIAL/NSN	4a. MILES	b. HOURS	c. ROUNDS FIRED	d. HOT STARTS	5. DATE	6. TYPE INSPECTION

7. APPLICABLE REFERENCE			
TM NUMBER	TM DATE	TM NUMBER	TM DATE

COLUMN a – Enter TM item number.

COLUMN b – Enter the applicable condition status symbol.

COLUMN c – Enter deficiencies and shortcomings.

COLUMN d – Show corrective action for deficiency or shortcoming listed in Column c.

COLUMN e – Individual ascertaining completed corrective action initial in this column.

STATUS SYMBOLS

"X" – Indicates a deficiency in the equipment that places it in an inoperable status.

CIRCLED "X" – Indicates a deficiency, however, the equipment may be operated under specific limitations as directed by higher authority or as prescribed locally, until corrective action can be accomplished.

HORIZONTAL DASH "(-)" – Indicates that a required inspection, component replacement, maintenance operation check, or test flight is due but has not been accomplished, or an overdue MWO has not been accomplished.

DIAGONAL "(/)" – Indicates a material defect other than a deficiency which must be corrected to increase efficiency or to make the item completely serviceable.

LAST NAME INITIAL IN BLACK, BLUE-BLACK INK, OR PENCIL - Indicates that a completely satisfactory condition exists.

FOR AIRCRAFT - Status symbols will be recorded in red.

ALL INSPECTIONS AND EQUIPMENT CONDITIONS RECORDED ON THIS FORM HAVE BEEN DETERMINED IN ACCORDANCE WITH DIAGNOSTIC PROCEDURES AND STANDARDS IN THE TM CITED HEREON.

8a. SIGNATURE *(Person(s) performing inspection)*	8b. TIME	9a. SIGNATURE *(Maintenance Supervisor)*	9b. TIME	10. MANHOURS REQUIRED

TM ITEM NO. a	STATUS b	DEFICIENCIES AND SHORTCOMINGS c	CORRECTIVE ACTION d	INITIAL WHEN CORRECTED e

Figure 6.4a DA Form 2404 (front)

TM ITEM NO. a	STATUS b	DEFICIENCIES AND SHORTCOMINGS c	CORRECTIVE ACTION d	INITIAL WHEN CORRECTED e

Figure 6.4b DA Form 2404 (back)

DON'T FORGET . . .

Unit level maintenance
is your
first line of defense!

The PMCS Workflow

Leaders such as you must carefully screen, train, license, and supervise Soldiers and others who are selected as system operators, drivers, and users. Unit equipment is mission essential—this requires that its operators be carefully trained, screened, and led.

The TM XX-10 and TM XX-20 series list the observation standards for all equipment. The elements of the process are established in DA PAM 750-8. An abbreviated version is described in paragraph 5-1*a* through *e*.

Operators and crews will use the equipment TM for *before*, *during*, and *after* operation PMCS. Soldiers perform operational checks and services before the equipment leaves the motor pool or other dispatch point.

During before-operations checks, Soldiers will correct all reparable faults. They will enter on the current DA Form 5988-E used during the before-operations checks any other faults not already recorded on a previously completed DA Form 5988-E or posted to DA Form 2408-14 (Uncorrected Fault Record).

Make operations checks during actual operation of the weapon system, vehicle, or other equipment, using the technical literature. If during the equipment mission no opportunity arises to correct a detected fault, the operator or crew chief must report the fault to the leadership at the unit dispatch point for corrective action.

During after-operations checks, the operator, crew, or mechanic will correct all known new faults, if possible. The commander's representative will decide whether to record any remaining faults on the uncorrected fault section of DA Form 5988-E or DA Form 2408-14. The nature of any uncorrected faults may dictate that the equipment not be cleared for future use or dispatch until the faults are corrected.

You and your Soldiers will use the TAMMS and STAMIS procedures identified in DA PAM 750-1 paragraph 4–4b to correct equipment faults discovered during PMCS operations and will update the DD Form 314 records as required.

Mechanics and supervisors should be licensed on every vehicle and piece of equipment they are associated with.

Don't forget: If it can't move, it does not pass the PMCS.

Your life and the lives of your Soldiers depend on a good PMCS. It can make the difference between whether you live or die.

The Not-Mission-Capable (NMC) Area Report of DA Forms 2406, 3266-1, and 1352

Commanders use information on materiel and unit readiness reports to analyze, predict, and make decisions on each unit's ability to perform its mission. These reports are completed during both peacetime and military operations and are useful only if they are timely, accurate, and complete. The materiel readiness of a unit reflects the ability of assigned equipment or systems to perform their missions.

For this reason, you should emphasize adherence to the Army Maintenance Standard (see AR 750-1). The most useful tools available to assess reading are:

- DA Form 2406
- using "screen shots" in Status Reports in accordance with AR 220-1
- DA Form 3266-1 (Army Missile Materiel Readiness Report)
- DA Form 1352 (Army Aircraft Inventory, Status, and Flying Time).

Commanders of units that perform equipment maintenance above the organizational level review each supported unit's readiness report. They then coordinate with the unit commander to prioritize maintenance requests and available resources. This helps achieve the highest equipment readiness possible for all supported units.

Readiness Reports

You should review readiness reports carefully and ask the following questions:

- Has any of the equipment listed been **not mission capable (NMC)** for an extended period—for example, three to five days?
- Is any item that has been NMC for more than seven days receiving intensive management? Require daily updates if necessary.
- If on work request for MAC-Code F level repairs, is your maintenance provider doing everything possible to assist you, including operation readiness float (ORF)?
- If repair parts have been ordered, is the interval between the date the item was identified and the date of the parts request greater than one working day? If so, what corrective action will you pursue?
- Are maintenance managers checking supply follow-ups and status cards?
- If an item was job ordered to support maintenance, was the interval between the date the item was identified as NMC and the date it went to support maintenance greater than two days? If the answer is "Yes," you must take corrective action within the organization. The item likely falls short of the Army Maintenance Standard in aspects other than the fault that makes it NMC.

not mission capable (NMC)

condition indicating that systems and equipment are not capable of performing their assigned missions because of maintenance requirements

> Unit status reports under the provisions of AR 220-1 can be reviewed only by personnel with the proper security clearance.

If it takes more than one calendar day between the time the maintenance provider organization notifies that a repaired item is ready for pickup, and the actual pickup of the serviceable item, you'll likely need to take internal organizational corrective action to meet the unit-response time requirements (see AR 750-1, chapter 3).

If there are external support delays, such as delays in the supply of repair parts, technical assistance, and other support, the local AMC logistics assistance officer or the appropriate AMC logistics assistance representative may provide needed information or assistance.

DA Form 2406

DA Form 2406 provides a standard format for manually reporting equipment condition.

AR 700-138 provides detailed instructions for preparing DA Form 2406. DA Form 2406 provides equipment status information for planning day-to-day operations on the maintenance workload and prioritizing work to produce the best readiness profile.

The most useful part of the DA Form 2406 for you is the back side of the completed form. Some commanders require their maintenance personnel to complete the back side daily, to ensure NMC equipment gets visibility and leadership attention. Equipment that is nonoperational for administrative or safety reasons is also noted in some commands.

The front side of the completed monthly DA Form 2406 is a historical report of equipment available over the reporting period, usually one month. You should review this completed form carefully to ensure its accuracy.

- Check firsthand to ensure your unit's quality of preparation. Check random DA Forms 5988-E/2404 against the corresponding DD Form 314.

- Using a specific model of equipment, compare the authorized quantity with the MTOE authorization, counting items that make up a system. If a part of a system is NMC, the whole system is NMC. Check AR 700-138, Table B-1, against MTOE equipment line-item numbers to determine if the correct number and types of systems are included in the report.

- Compare the on-hand quantity of a single model of equipment to the number of copies of DD Form 314, noting substitute line items. Check the report period to verify the "possible" days for reporting purposes.

- Total the nonavailable days taken from the copies of the DD Form 314, and subtract this number from the possible days to verify available days.

- Ensure that the nonavailable days are divided correctly into supply and maintenance categories for both MAC-Code O and MAC-Code F (formerly unit and support) maintenance.

- Compare these numbers to the daily DA Form 2406 (DA Forms 5409/5410, if not using STAMIS/SAMS-E), and file copies of DA Form 2407.

- Provide explanations for any differences. In modularized organizations, readiness information is available from the internal MAC-Code F maintenance element. If the MAC-Code F maintenance provider is outside your organization and uses SAMS/SAMS-E, request automated reports from the provider to double-check DA Form 2406. Ask for the following:
  - Equipment nonoperational over the number of days covered by the report, compiled for each unit (reparable items by unit report, production control number (PCN) AHO 003)
  - Equipment nonoperational over the number of days covered by the report, compiled for each battalion (reparable items by battalion report, PCN AHO 026)
  - Customer work order reconciliation report, PCN AHN 004.

- The equipment nonoperational reports can be requested to cover a company-sized unit or battalion for as many days as needed. The report period for a DA Form 2406 should provide data to match the DA Form 2406 back side. The customer reconciliation report lists all work orders the maintenance provider organization has in an "open" status for the unit/organization. It contains NMC-S and NMC-maintenance (NMC-M) time. If you discover discrepancies, seek explanations.

Examine closely the readiness profiles in the report and identify items that degrade readiness.

The broad factors and indicators that you should review include the following:

- Unit maintenance performance during the most recent readiness exercise or the Army Training Evaluation Program
- Availability of maintenance leadership and skills
- Maintenance training requirements and shortfalls. If help is needed, note this on the unit's status report. Commanders decide the overall readiness status based on their observations, statistical data, and informed judgment.

MATERIEL CONDITION STATUS REPORT

For use of this form, see AR 700-138; the proponent agency is DCS, G-4.

Requirement Control Symbol CSGLD-1042(R4)

1. PERIOD OF REPORT		2. DATE PREPARED	3. UTILIZATION CODE	4a. PAGE NO	4b. NO PAGES
FROM	TO				

5. TO (Address including ZIP Code)

6. FROM (Address including ZIP Code)

7. UNIT IDENTIFICATION CODE

8. TOE NO

9. AVAILABILITY STATUS (Itemized)

a. SEQ NO.	b. NOMENCLATURE			c. ECC LIN	d. DENSITY		e. EQUIPMENT AVAILABILITY						f. FOR FIELD USE ONLY				
	(1) NOUN	(2) EOS	(3) MODEL		(1) AUTH QTY	(2) ON-HAND QTY	(1) POSSIBLE DAYS	(2) AVAILABLE DAYS	(3) NONAVAILABILITY DAYS					(1) REQ QTY	(2) REQ DAYS	(3) FMC	(4) ER	(5) ERC
									(a) ORG		(b) SPT							
									S	M	S	M					

Figure 6.5a   DA Form 2406 (front)

10. NONAVAILABILITY STATUS *(itemized)*

a. SEQ NO.	b. NOMENCLATURE		c. REGISTRATION OR SERIAL NO.	d. NON-AVAILABILITY REASON	e. DATE NON-AVAILABLE	f. DATE ADMITTED TO SHOP		g. SUPPORT SHOP JOB OR RON NO AND DODAAC	h. REMARKS NSN OR PART NO
	(1) NOUN	(2) MODEL				(1) ORG	(2) MAINTENANCE SUPPORT		

11. REMARKS

12a. VERIFIED BY *(Signature)*

12b. DATE

*NOTE: Indicate reason for nonavailability as follows: A - Modification; B - Parts; C - Malfunction; D - Support maintenance.

Figure 6.5b DA Form 2406 (back)

ARMY MISSILE MATERIEL READINESS REPORT
For use of this form, see AR 700-138; the proponent agency is DCS, G-4

1. DO NOT WRITE IN THIS SPACE	REQUIREMENT CONTROL SYMBOL CSGLD-1864 (R1)

2. TO *(Include ZIP Code)*	3. FROM *(Include ZIP Code)*		
		4. UIC	
		5. PERIOD ENDING	6. DODAAC
		7. DSN	

8. PART I - SYSTEM OPERATIONAL DATA

a. FMC %	b. NMCS %	c. NMCM %	NMCS	NMCM
d. WEAPON SYSTEM	e. REQ	f. AUTH	g. ON HAND	h. POSSIBLE HOURS/DAYS	i. MISSION CAPABLE HOURS/DAYS	j. ORG	k. SUP	l. ORG	m. SUP

9. PART II - SYSTEM COMPONENT OPERATIONAL DATA

LINE	ITEM a.	SERIAL NO. b.	POSSIBLE HOURS/DAYS c.	FMC HOURS/DAYS d.	NMCS ORG e.	NMCS SUP f.	NMCM ORG g.	NMCM SUP h.
1								
2								
3								
4								
5								
6								
7								
8								
9								
10								
11								
12								
13								
14								
15								
16								

10. NAME AND GRADE OF AUTHENTICATING OFFICER *(Type or print)*	11. SIGNATURE

Figure 6.6a DA Form 3266-1 (front)

12. PART III - NOT MISSION CAPABLE STATUS ITEMS				
ITEM a.	SERIAL NUMBER b.	DATE NON-AVAIL. c.	DS/GS JOB ORDER NO. OR DOCUMENT NO. *(Include DODAAC)* d.	MALFUNCTION OR PART NO. e.

13. COMMANDER'S READINESS IMPACT STATEMENT

Figure 6.6b DA Form 3266-1 (back)

ARMY AIRCRAFT INVENTORY, STATUS AND FLYING TIME
For use of this form, see AR 700-138; the proponent agency is DCS, G-4

1. PERIOD ENDING	2. PAGE NO.	3. NO. OF PAGES
REQUIREMENT CONTROL SYMBOL CSGLD-1837(R1)

4. ORGANIZATION

5. TELEPHONE (Comm/DSN)

6. UNIT IDENTIFICATION CODE

7. (Do not write in this space)

8. POST, CAMP, STATION

9. COMMAND

10. SUMMARY DATA

MISSION DESIGN SERIES a	SERIAL NUMBER b	ASSIGNMENT AND FUNCTIONAL CODE c	HRS. ON HAND DURING REPORT PERIOD d	MISSION CAPABLE			NOT MISSION CAPABLE				HOURS FLOWN DURING MONTH k	NUMBER OF LANDINGS / TOUGHDOWN AUTO-ROTATIONS l	GAINED OR LOST m
				FMC e	FMC PMCM f PMCS	NMCS g	DEPOT h	NMCM AVIM i	AVIM i	AVUM j			

11. TYPED OR PRINTED NAME, GRADE, AND POSITION OF AUTHENTICATING OFFICER

12. SIGNATURE

Figure 6.7 DA Form 1352

TABLE 6.3A			List of Ground Equipment for DA Form 2406			

ECC	LIN	EIC	Nomenclature	Abbreviation	Model No.	NSN
JA	A06352	IPR	Aviators Night Vision Imaging System	ANVIS	AVS6V1	5855011384749
JA	A06420	IPQ	Aviators Night Vision Imaging System	ANVIS	AVS6V2	5855011384748
GZ	A10769	ATB	Adapter Hardware FVS Peculiar	ADPT HDWR	STEFVS	4910011354379
GZ	A10837	ATE	Adapter Hardware M1 Peculiar	ADPT HDWR	STEM1	4910011422640
JS	A27159	JPX	Air Traffic Control Facility	ATC FAC	TSQ97	5895001378548
JS	A27624*	JP3	Air Traffic Control Central	ATC CEN	TSW7A	5895010181246
JS	A28833	JP9	Aircraft Control Central	AC CEN	TSQ70	5895001681576
		JPY			TSQ70A	5895001681577
JP	A41666*	IYB	Radar Set	RDR ST	TPQ37V1	5840010434258
		IYD			TPQ37V2	5840010845374
		IYK			TPQ37V3	5840011869125
		IYJ			TPQ37V4	5840011854243
		IYG			TPQ37V5	5840012705101
		IYF			TPQ37V6	5840012705100
		IT7			TPQ37V8	5840014003218
QW	A48430*	5AP	Alarm, Biological Agent Automatic Integrated System	ALARM BIO AGENT	M31	6665013926191
QW	A48498*	5AQ	Alarm, Biological Agent, Automatic	ALARM BIO	M31A1	6665014362309
OC	A55656	8HD	Analyzer	ANAL CL	QBCII	6630013165085
KC	A56243	B9A	Analyzer Set Engine Portable	ANAL ST	STEICEPM	4910001242554
		B9C			STEICR	4910012226589
OA	A62773	8BA	Anesthesia App, Nitrous	ANES AP NI	885A	6515011858446
		8BE			885	6515010034133
OB	A82942	8HJ	ANALYZER CHEMICAL	ANAL CC	DT60	6630013769823
OR	A84549	8HE	Analyzer Sodium, Potassium	ANAL SP	614	6630013008711
GM	A93125*	ALB	Armored Reconnaissance Airborne Assault Vehicle 152MM	ARAAV	M551A1	2350001405151
LB	B25476	XJI	Boat Bridge Erect., Hydro Jet	BOAT BRDG	MK1	1940011055728
		XJJ			MK2	1940012189165
NL	B31098*	ARF	Bridge (AVLB)	BRDG AVLB	MLC70	5420013903933
OB	B32900	8HI	Analyzer Blood Gas	ANAL BL	4300	6630013648555
QE	B43663	ZKP	Bath Unit Portable	BATH UT	SH63LP	4510010163332
		ZKR			8SH60LP	4510010165914
		ZKS			YS49279LP	4510010165915
		ZKT			SPE41LP	4510010217421
		ZKU			8SH70YSLP	4510010229620
		ZKV			8SH1LP	4510010272123
		ZKX			YS74LP	4510010745177
		ZKZ			YS8SH76LP	4510010802402
		ZK4			PORT9SH	4510011394973

*Denotes that items will be reported as systems.

TABLE 6.3B			List of Ground Equipment for DA Form 2406, continued			

ECC	LIN	EIC	Nomenclature	Abbreviation	Model No.	NSN
JH	B51098	JPN	Beacon Set Radio	BCN ST RDO	TRN30V1	5825004054510
JH	B51099	JPP	Beacon Set Radio	BCN ST RDO	TRN30V2	5825004231654
HX	B83002	DVY	Bed Cargo Demountable PLS	BD CGO DMT	M1077	3990013077676
		DV2		BD CGO DMT	M1077	3990014061340
		DV7		BD CGO DMT	CROP	3990014422751
LB	B83582	XJA	Boat, Bridge Erection Propeller Propulsion	BOAT BRDG	T15	1940003554469
		XJD			MDL27	1940005260207
		XJC			DSLENG	1940004170526
		XJE			LONESTAR	1940005677898
		XJF			MRNTMD27	1940007106649
		XJG			HIWAY	1940008094472
		XJH			HP127C	1940009150079
GR	C00255	BXE	Carrier Ambulance 1½T	CARR AMB	M1066	2350012836215
GL	C00384*	AP6	Carrier Air Defense	CARR AIR D	M6ODS	2350014480368
GR	C10908	AEW	Carrier, Ammo, Tracked	FAASV	M992	2350011104660
		AE6			M992A1	2350013523021
		AKA			M992A2	2350013689500
GB	C10990*	AE4	Carrier 120MM Mortar, Self-Propelled, Armored	CARR MTR	M1064	2350013383116
		AE8			M1064A3	2350013696082
GQ	C11158	AE5	Carrier Armored, Command Post, Full Tracked	CARR CP	M1068	2350013545657
		AFC			M1068A3	2350013696086
GR	C11280	BXA	Carrier, Cargo, Tracked 1.5T	CARR CGO	M973	2350011329099
		BXB			M973A1	2350012816451
GR	C11651	BXD	Carrier Command Communication Vehicle	CARR CMD	M1065	2350012818324
GW	C12815*	AES	Carrier, Smoke, Gen FT, AR	CARR SM GE	M1059	2350012030188
		AFA			M1059A3	2350013696083
LY	C14504	WBP	Causeway System Floating	CAUSEWAY	Floating	1945012187268
LY	C14572	WBQ	Roro Discharge Facility	CAUSEWAY	ORODF	1945012192109
GR	C16921	BXC	Carrier Cargo Flatbed, 2T	CARR FB	M1067	2350012816450
JY	C17936*	GE5	Field Artillery Computer Set	FD ART COMP ST	ANGYG3V1	1220014524303
JY	C17832	QT2	Computer Set, Digital	COMP ST DIG	OL587TYQ	7010014204985
JY	C18072*	GE4	Field Artillery Computer Set	FD ART ST	ANGYG3V4	1220014523567
GL	C18234*	AEY	Carrier Personnel, Full Tracked	CARR PERS	M113A3	2350012197577

*Denotes that items will be reported as systems.

TABLE 6.3C			List of Ground Equipment for DA Form 2406, continued			

ECC	LIN	EIC	Nomenclature	Abbreviation	Model No.	NSN
JY	C18242	QTV	Computer Set, Digital	COMP ST DIG	OL602TYQ	7010014204982
JY	C18310	QTU			OL601TYQ	7010014204984
JY	C18344	QTJ			OL605TYQ	7010014204965
JY	C18412	QTK			OL606TYQ	7010014204964
JY	C18446	QTC			OL582TYQ	7010014194989
JY	C18480	QTL			OL607TYQ	7010014204963
JY	C18514	QTD			OL583TYQ	7010014194987
JY	C18548	QTN			OL609TYQ	7010014204979
JY	C18582	QTA			OL584TYQ	7010014194988
JY	C18684	QTT			OL604TYQ	7010014204981
JY	C18718	QTR			OL591TYQ	7010014204976
NL	C20414	ARA	Bridge Armor Veh Launch Scissor TY CL 60 Alum 60 FT Span	AVLS	AVLSC60	5420005229599
NK	C22058	XHI	Bridge Erect Set Fix	BDGE ER ST	97CLEO40	5420005303785
NK	C22126	XHA	Bridge Erect Set Fix Medium Girder Bridge	MGB	97CLE53	5420001723519
NK	C22811	XHB	Bridge Fixed, Medium Girder Bridge	MGB	97CLE52	5420012723520
NK	C23017	XHH	Bridge Fixed, HWY	BDGE FIX	MILB11844	5420005303784
NO	C25072	XJK	Bridge Floating HWY Alum Deck	BDGE FLTG	97CLE35	5420001714519
NO	C25346	XJU	Bridge Floating HW 135 ft.	BRDG FL HW	CL60135	5420000599082
NO	C25757	XJR	Bridge Floating Raft Sect Light Tact	BDGE FLTG	97CLE42	5420005424719
NK	C26305	XJT	Bridge Erect Set Floating Bridge	BDGE ER ST	CL60	5420008924596
JY	C27007*	GE6	Field Artillery Computer Set	FD ART COM ST	ANGYG3V2	1220014524304
NK	C27309	XHC	Reinforcement Set, Medium Girder Bridge	REINF ST	97CLE56	5420011391503
JY	C27823	QTM	Computer Set, Digital	COMP ST DIG	OL608TYQ	7010014204962
JS	C28728*	JQ3	Central Communication	CENT COMM	ANTSQ190v4	5895013995915
JH	C30675*	L6H	Countermeasures Set	CTRMSR	TLQ17AV3	5865012752137
JY	C35900*	L3H	Communications Ctl St Comm	CTL	TSQ183	5895013696166
		LDR			TSQ183B	5895014422087
JY	C36104*	LE2	Communications Ctl St	COMM CTL	TSQ184B	5895013875801
		LEK			TSQ184E	5895014422095
NF	C36151	EKY	Crane, Wheel Mtd, HYD 7½ Ton	CRANE MTD	LRT110	3810011650646
PK	C38874	DSA	Crane Truck Mtd, 140 Ton Container	CRANE MTD	ACN21086	3810010279254
		DSF			HC238A	3950011109224
NF	C39398	EKG	Crane, Wheel Mtd, HYD, Rough Terrain	CRANE MTD	RT875	3810012052716
JS	C41061*	HN8	Central Message Switching Automatic	CEN MSG SA	TYC39A	5805013635118
		HLZ			TYC39V1	5805011231851
		HN7			TYC39V5	5805011523068

*Denotes that items will be reported as systems.

TABLE 6.3D			List of Ground Equipment for DA Form 2406, continued			

ECC	LIN	EIC	Nomenclature	Abbreviation	Model No.	NSN
JC	C41311*	HNC	Central Office Telephone, Automatic	COTA	TTC39AV1	5805012419710
		HN5			TTC39D	5805013153751
		JFX			TTC39EV1	5805013862830
JS	C59125*	GB5	Communication Sys	Comm Sys	TSQ198	5895013881454
GL	C76335*	APB	Fighting Vehicle, Cavalry	CFV	M3	2350010492695
JY	C77687	GE7	Computer, Fire Control	COMP FI CON	COMPANP SG8V1	1270013765614
		LDJ			SG8V1	7035014449249
JY	C78486	QTZ	Computer Set, Digital	COMP ST DIG	OL586TYQ	7010014126730
	C78554	QT4			OL589TYQ	7010014204986
	C78759	JFV			ANTYQ85	7010014500332
JC	C78793*	HLN	Central Office Telephone, Automatic	COTA	TTC41V2	5805010288394
JY	C78827	QTS	Computer Set, Digital	COMP ST DIG	OL603TYQ	7010014204983
JC	C78861*	HLL	Central Office Telephone, Automatic	COTA	TTC41V3	5805010288392
JY	C78895	QTB	Computer Set, Digital	COMP ST DIG	OL585TYQ	7010014194990
JC	C78929*	HLT	Central Office Telephone, Automatic	COTA	TTC41V4	5805010448869
SA	C82833	YTZ	Camera Section, Topographic Reproduction Set	CAMERA SCT	97CLE221	3610003444706
		YT2			TEADTSS22	3610011051694
QX	C84541*	V4H	Container Assy Ref	Ref Cont	SC200	8110010157039
		ZVT			SC210	8145013379996
JS	C89935*	JQ2	Central Communications	CEN COMM	TSQ190V3	5895013935224
	C90003*	JQY			TSQ190V1	5895013787993
	C90071*	JQZ			TSQ190V2	5895013790125
JY	C90531*	L3G	Communications Control Set	COMM CTL	TSQ182	5895013696170
		LDK			TSQ182A	5895014422098
JY	C90599*	GAU	Communications Control Set	COMM CTL	TSQ183A	5895013875792
		LDS			TSQ183C	5895014422096
JY	C90667*	L3J	Communications Control Set	COMM CTL	TSQ184	5895013696167
		LEB			TSQ184C	5895014422094
JY	C90735*	JQY	Communication Control Set	COMM CTL	TSQ184A	5895013875620
		LEC			TSQ185Dd	5895014417285
JC	C91132	LMB	Communications Terminal	COMM TR	TRC179V1	5895011560411
JY	D10281*	GE8	Digital Topographic Support System	DTSS LIGHT	ANTYQ67V1	6675014248516
GB	D10741*	AER	Carrier Mortar, Self Propelled 107MM	CARR MRTR	M106A2	2350010696931
GR	D11049	AEU	Carrier, Cargo Full Tracked 6 Ton	CARR CGO	M548A1	2350010969356
		AE9			M548A3	2350013696081
JY	D11248*	GE3	Digital Topographic Support System	DTSS HEAVY	ANTYQ48A	6675014422105

*Denotes that items will be reported as systems.

TABLE 6.3E			List of Ground Equipment for DA Form 2406, continued			

ECC	LIN	EIC	Nomenclature	Abbreviation	Model No.	NSN
GQ	D11538*	AEQ	Carrier, Command Post: Light Tracked	CARR CP	M577A2	2350010684089
		AE7			M577A3	2350013696085
GL	D12087*	AEN	Carrier, Personnel Full Tracked AR	CARR PER	M113A2	2350010684077
JC	D18673	GB3	Dismounted Extension Switch	DES	TTC51	5895013498065
JR	D18923	IYL	Radio, Dismounted Line of Sight, Multichannel	RDO DLOS	TRC198V2	5820013499240
JY	D31557	HP4	Data Display Group, Gun Direction	DDGGD	OD144V1	7025011342329
JY	D31625	HQH	Data Display Group, Gun Direction	DDGGD	OD144V2	7025011343218
JY	D31693	HQJ	Data Display Group, Gun Direction	DDGGD	OD144V3	7025011343219
JY	D40782	GLJ	Digital Message Device Group	DIG MSG DV	OA8990P	5820011023921
JY	D78075*	HPS	Data Processing Systems Automated	DP SYS	MYQ4	7010010906819
JY	D78325*	HYB	Data Processing Systems Automated	DP SYS	MYQ4A	7010011585397
QM	D82404*	5FC	Decontaminating App Pwr Drvn LT WT	DECON APP	AE32U8	4230011538660
		5FE			M17	4230012518702
		5FF			M17A1	4230013035225
		5FG			M17A2	4230013461778
		5FH			M17A3	4230013463122
OE	D86072	8BF	Defibrillator ECG Monitor/Recorder	DEF ECG	MRL90	6515011350840
		8BJ			43110MC	6515012911199
		8BQ			LifePack 10	6515013896740
NJ	D95754	ZJO	Drilling Machine, Well Truck Mounted	DR MACH		3820011785057
OR	E17489	8EI	Edging Machine Ophthalmic Lens E	DG MACH	All models	6540001165780
GG	E56578*	ABF	Combat Engineer Vehicle Full Tracked	CBT EN VEH	M728	2350007951797
JH	E59831	LHJ	Communications Central	COMM CEN	TSC38B	5895001681487
NV	E61618	EXB	Compactor, High Speed Tamping, Self-Propelled	CMPTR HS	K300	3805010244064
OG	E67355	8CA	Compressor Dehydrator Dental	COMP DEN	M5SERIES	6520001391246
		8CC			CN60358	6520012422375
		8CK			PAC6	7652013984613
QC	E72393	ZPV	Compressor Unit, Rotary, 125 CFM 100 psi skid Mtd	COMPR RTY	6M125	4310010437604
		ZQA			125GC40MS3	4310006910877
		ZQB			GER125	4310008189824
QC	E72804	DWT	Compressor Unit, Rotary, 210 CFM 100 PSI, Air Trlr Mtd	COMPR RTY	250WDMH268	4310011583262
NF	F39378	EKC	Crane Wheel Mounted 20 Ton	CR WHL 20T	M320RT	3810002751167
GL	F40307*	ALE	Fighting Vehicle Infantry	IFV	M2A1	2350011791027
GL	F40375*	ALG	Fighting Vehicle Infantry	IFV	M2A2	2350012487619
		APE			M2A2WODS	2350014059886

*Denotes that items will be reported as systems.

TABLE 6.3F			List of Ground Equipment for DA Form 2406, continued			

ECC	LIN	EIC	Nomenclature	Abbreviation	Model No.	NSN
NF	F40474	EMK	Crane Shovel, Crawler Mtd 40 Ton	CR SHVL	PH5060	3810011458288
QJ	F42612	ZIV	Forward Area Water Point Supply System	FAWPSS	FAWPSS	4320011101993
		ZFW			90952	
JC	F43336*	GB2	Force Entry Switch	FES	TC50	5895013498064
NF	F43429	ELA	Crane Truck Mtd HYD 25 Ton CAT (CCE)	CR TK 25T	MT250	3810000182021
		ELH			TMS3005	3810010549779
NA	F49399	EUT	Crush and Screen Plant	CR SCN PLT	75TPH	3820007256462
NA	F49673*	EWL	Crush Screen & Wash Plant	DSL ELEC	225TPH	3820005278577
		E5G			AN WA	3820014355177
JY	F55539*	GDM	Fire Control Sys FA	Fire CTL FA	ANGYK37V1	1230013598522
JY	F55750*	P9	Fire Direction Center	FDCA	OA8390	7010010177040
		HZD		FDCA	OA8390BV2	7010012518585
JX	F57463	HP2	Fire Support Digital Device	FSDMD	PSG5	7025011256796
GL	F60462*	ALF	Cavalry Fighting Vehicle	CFV	M3A1	2350011791028
GL	F60530*	ALH	Cavalry Fighting Vehicle	CFV	M3A2	2350012487620
		APF			M3A2WODS	2350014059887
QM	F81880*	5FB	Decontaminating Apparatus, Power Driven Skid Mtd	DCON APPR	M12A1	4230009269488
OF	F95601	8CB	Dental Operating Treatment Unit, Field	DTL OP UT	ALL MODELS	6520001407663
		8CD			G283	6520012052349
		8CJ			36009900	
		8CH			FUS336	6520012724531
QB	G11966	VG2	Generator Set, Dsl, 5KW, 60HZ, Skid Mtd	GEN ST SM	MEP802A	6115012747387
QB	G12034	VG7	Generator Set, Dsl, 60KW, 50/60HZ, Skid Mtd	GEN ST SM	MEP806A	6115012747390
QB	G12102	VN2	Generator Set, Dsl, 5KW, 400HZ, Skid Mtd	GEN ST SM	MEP812A	6115012747391
QB	G12170	VG4	Generator Set, Dsl, 15KW, 50/60HZ, Skid Mtd	GEN ST SM	MEP804A	6115012747388
QB	G12238	VN4	Generator Set, Dsl, 15KW, 400HZ, Skid Mtd	GEN ST SM	MEP814A	6115012747393
QB	G17460	VNB	Generator Set, Dsl, 60KW, 400HZ, Trl Mtd	GEN ST TM	PU806	6115013172133
QB	G18052	VN6	Generator Set, Dsl, 60KW, 400HZ, Skid Mtd	GEN ST SM	MEP816A	6115012747395
QB	G18358	VG6	Generator Set, Dsl, 3KW, 60HZ, Skid Mtd	GEN ST SM	MEP831	6115012853012
QB	G35851	VD4	Generator Set, Dsl, Trl Mtd	GEN ST TM	PU803	6115013172136

*Denotes that items will be reported as systems.

TABLE 6.3G			List of Ground Equipment for DA Form 2406, continued			

ECC	LIN	EIC	Nomenclature	Abbreviation	Model No.	NSN
QB	G35919	VMZ	Generator Set, Dsl, Trl Mtd	GEN ST TM	PU804	6115013172135
QB	G37273	VJW	Generator Set, DSL, 5HZ, 60HZ, Mtd on M116	GEN ST TM	PU751M	6115000331373
QB	G40744	VJB	Generator Set, DSL, 10KW, 60HZ, Mtd on M116	GEN ST TM	PU753M	6115000331389
QB	G42170	VK5				
		VNC	Generator Set, 10KW, 60HZ, Mtd on M116A2	GEN ST TM	PU798	6115013199032
					PU798A	6115014133818
QB	G42238	VKK	Generator Set, 5KW, 60HZ, Mtd on M116A2	GEN ST TM	PU797	6115013320741
		VND			PU797A	6115014133820
GX	G51840*	5CD	Generator Set, Smoke	GEN ST SMK	M157120GT	1040012060147
		5CE			M15780GT	1040012935496
		5CI			M157A28OD	1040014068923
		5CH			M157A212OD	1040014067401
QB	G53403	VK4	Generator Set, 10KW, 400HZ, Mtd on M116A2	GEN ST TM	PU799	6115013134283
		VDW			PU799A	6115014133819
QB	G53778	VD3	Generator Set, Dsl, Trl Mtd	GEN ST TM	PU802	6115013172138
QB	G54041	VGV	Generator Set, Dsl, 3KW, 60HZ, Skid Mtd	GEN ST SM	MEP701A	6115012345966
		VGW			MEP016B	6115011504140
GX	G58151*	5CF	Generator, Smoke, MECH	GEN ST SMK	M356	1040013801400
QB	G74575	VG5	Generator Set, Dsl, 30KW, 50/60HZ, Skid Mtd	GEN ST SM	MEP805A	6115012747389
QB	G74643	VN5	Generator Set, Dsl, 30KW, 400HZ, Skid Mtd	GEN ST SM	MEP815A	6115012747394
QB	G74711	VG3	Generator Set, 10kw Dsl	GEN ST TM	MEP803A	6115012755061
QB	G74779	VN3	Generator Set, Dsl, 10KW, 400HZ, Skid Mtd	GEN ST SM	MEP813A	6115012747392
NE	G74783	EHF	Grader Road Motorized DED	GRDR ROAD	130G	3805011504795
QB	G78203	VMY	Generator Set, 15KW, 400HZ, Trl Mtd	GEN ST TM	PU800	6115013172137
QB	G78306	VF3	Generator Set, Dsl, 60KW, 50/60HZ, Trl Mtd	GEN ST TM	PU805	6115013172134
GX	G87229*	5CG	Mech Smoke Generator	GEN SMK	M58	1040013801400
JR	H35404	GGE	High Frequency Radio Set	RDO ST HF	GRC193A	5820011334195
		GGT			GRC193BV1	5280012629546
LK	H38787	XJO	Ferry Conversion Set Raft, Inf Spt	FERRY	97CLE05	5420002729267
VC	H56391	ZML	Fire Fighting Equipment Set: Truck Mounted	2500L	FFES MTD	4210011522699
		ZMN			CL530	4210002028076

*Denotes that items will be reported as systems.

| TABLE 6.3H | | | List of Ground Equipment for DA Form 2406, continued | | | |

ECC	LIN	EIC	Nomenclature	Abbreviation	Model No.	NSN
DA	H57505*	3FA	Howitzer, Light Towed	HOW LT TWD	M119	1015012480859
		3WC			M119A1	1015013081872
					105MM	
GA	H57642*	3FC	Howitzer, Medium Self-Propelled	HOW MED SP	M109A6	2350013050028
JS	H76352*	JQC	Flight Coordination Central	FLT CEN	TSC61LP	5895001681573
		JQB			TSC61ALP	5895000113878
		JP4			TSC61BLP	5895010573968
QH	H94824	ZAG	Forward Area Refueling Equipment	FARE	FARE	4930001333041
		ZA4			LPIF0500	4930013018201
QH	J04717*	ZAH	Fuel System Supply Pt, Ptbl, 600,000 Gallon	FSSP	FSSP	4930001425313
QB	J30093	VEP	Generating Unit, DSL, 750 KW, 60HZ		MEP208A	6115004505881
		VFK			S6660	6115005591449
		VC8			S6832	
EY	J30492*	5CA	Generator: Smoke Mechanical Pulse Jet	GEN SMK	M3A3	1040005873618
		5CB			M3A4	1040011439506
QB	J35492	VCN	Generator Set, DSL, 15KW, 60HZ	GEN ST TM	PU405AM	6115003949577
QB	J35629	VEM	Generator Set, DSL, 60KW, 60HZ	GEN ST TM	PU650BG	6115002581622
QB	J35680	VLM	Generator Set, DSL, 60KW, 400HZ	GEN ST TM	PU707AM	6115003949573
QB	J35801	VDT	Generator Set, DSL, 100KW, 60HZ	GEN ST TM	PU495BG	6115011340165
QB	J35813	VJF	Generator Set, DSL, 5KW, 50HZ	GEN ST	MEP002A	6115004651044
QB	J35825	VJE	Generator Set, DSL, 10KW, 60HZ	GEN ST	MEP003A	6115004651030
		VJU			1480021	6115009373523
QB	J35835	VCD	Generator Set, DSL, 15KW	GEN ST	MEP004A	6115001181241
		VDC			15H18Z	6115005916866
		VDD			10327BA	6115006069693
		VDG			015H18M	6115006279031
		VDH			151815WW	6115006535634
		VDN			151815WA	6115008174919
QB	J36006	VLF	Generator Set, DSL, 15KW, 400HZ	GEN ST	MEP113A	6115001181244
QB	J36109	VCC	Generator Set, DSL, 30KW, 60HZ	GEN ST	MEP005A	6115001181240
QB	J36383	VCM	Generator Set, DSL, 30KW, 60HZ	GEN ST TM	PU406BM	6115003949576
QB	J36725	VLG	Generator Set, DSL, 30KW, 400HZ	GEN ST	MEP114A	6115001181248
QB	J38506	VLH	Generator Set, DSL, 60KW, 400HZ	GEN ST	MEP115A	6115001181253
QB	J38712	VCG	Generator Set, DSL, 100KW, 60HZ	GEN ST	MEP007A	6115001339101
		VDS			MEP007B	6115010366374
		VDL			4115	6115007922541
QB	J43027	VL8	Generator Set, Gas, 0.5KW, 400HZ	GEN ST	MEP019A	6115009407862

*Denotes that items will be reported as systems.

TABLE 6.31			List of Ground Equipment for DA Form 2406, continued			
ECC	LIN	EIC	Nomenclature	Abbreviation	Model No.	NSN
QB	J43918	VGC	Generator Set, Gas, 1.5KW, 60HZ	GEN ST	KK15M25	6115005916867
		VGF			1536S2A016	6115007749342
		VGI			CEO15AC	6115008878644
		VGJ			MEP015A	6115008891446
QB	J44055	VHA	Generator Set, Gas, 1.5KW, 28V DC	GEN ST	MEP025A	6115000178236
		VHD			GEMTRCE15L	6115006466122
		VHF			1528T2A016	6115008492323
QB	J45699	VGA	Generator Set, Gas, 3KW, 60HZ AC	GEN ST	MEP016A	6115000178237
		VGO			MEP016C	6115011433311
QB	J45836	VLA	Generator Set, Gas, 3KW, 400HZ AC	GEN ST	MEP021A	6115000178238
		VMT			MEP021C	6115011757321
QB	J46110	VHB	Generator Set, Gas, 3KW, 28V DC	GEN ST	MEP026A	6115000178239
		VHJ			MEP026C	6115011757320
QB	J46252	VGH	Power Unit, 3KW, 60HZ AC	GEN ST	PU PU625G	6115008733915
QB	J46384	VGE	Power Unit, 3KW, 60AZ AC	GEN ST	PU PU617M	6115007386335
QB	J47068	VJA	Generator Set, Gas, 5KW, 60HZ AC	GEN ST	MEP017A	6115000178240
QB	J47617	VJO	Power Unit, 5KW, 60HZ AC	GEN ST	PU PU620M	6115007386340
QB	J49398	VJT	Generator Set, Gas, 10KW, 60HZ AC	GEN ST	MEP018A	6115008891447
NE	J74852	EJG	Grader, Road, Motorized	GRDR RD	12	3805001974184
		EJM			116	3805002211802
		EJN			550	3805002239030
NE	J74886	EHL	Grader, Road, Motorized DSL	GRDR RD	CAT112FWR	3805010290140
		EHP			130GS	3805011267895
		EJH			130GSCE	3805012518252
NE	J74920	EHN	Grader, Road, Motorized	GRDR RD	130GNS	3805011267894
		EJJ			130GNSCE	3805012520128
GL	J81750*	APA	Fighting Vehicle, Infantry	IFV	M2	2350010485920
GA	K56981	3E5	Howitzer Hvy Sp 8 In	HOW HV SP	M110A1	2350010133914
		3E4			M110	2350004396243
		3E3			M110A2	2350010414590
DA	K57392*	3EA	Howitzer, TWD LT	HOW LT TWD	M102	1015000868164
		3EB			M101LT	1015003229728
		3EC			M101A1LT	1015003229752
GA	K57667	3ER	Howitzer, Medium, Self-Propelled: 155MM	HOW MD SP	M109	2350004408811
		3EZ			M109A2	2350010310586
		3E2			M109A3SP	2350010318851
		3E8			M109A4	2350012775770
		3E7			M109A5	2350012811719

*Denotes that items will be reported as systems.

TABLE 6.3J			List of Ground Equipment for DA Form 2406, continued			

ECC	LIN	EIC	Nomenclature	Abbreviation	Model No.	NSN
DA	K57803*	3EG	Howitzer Med TWD	HOW MD TWD	M114	1025003229755
		3EH			M114A1	1025003229768
		3EK			M114A2	1025010259857
DA	K57821*	3EL	Howitzer, Medium, Towed: 155MM	HOW MD TWD	M198	1025010266648
QS	K90188	BMW	Instrument Repair Shop, Truck Mounted	REP SHP TM	M185A3	4940000771638
LM	K97376	XMB	Interior Bay Bridge Floating	IBBF	IBBF	5420000715322
JH	L12374	L6I	Lightweight Man Trspbl Radio Directional Finding System	LMRDFS	PRD12	5825012986961
JS	L36402*	JQA	Landing Control Central	LDG CT CEN	TSQ71ALP	5895000040973
		JP5			TSQ71BLP	5895010928074
LD	L36739	WAE	Landing Craft, Mechanized: 69FT	LCM	LCM8	1905002671097
		WAS			LCM8MOD1	1905009356057
		WGC			LCM8MOD1SL	1905012842647
		WGD			LC08	1905012842648
LD	L36876	WAA	Landing Craft, Utility: 115FT	LCU	1646GEN	1905001685764
		WAV			1646MAR	1905010091056
LD	L36989	WBS	Landing Craft Util Roll On Roll Off	LCU	MDL2000	1905011541191
GK	L43664*	ARC	Launch Tank Chassis, Transporting, 60FT Bridge	LNCH TNK C	M60	5420008892020
		ARE			M48A5	5420010766096
JR	L61778	IYM	Radio, LF, Line of Sight, Multichannel	RDO	MC TRC198V1	5820013499241
DE	L67342*	556	Launcher, Mine Clearing Line Charge, Trailer Mounted	LCHR MCL	MK155	1055012035883
		59A			MK155M1	1055012812770
		5UJ			MK155M2	1055013406084
		5UK			MK155M3	1055013273106
LL	L67508	WAN	Lighter, Amphibious: Self-Propelled Diesel	LGTR AMPH	LARCLX	1930003922981
JS	L67964	HYD	Lightweight Digital Facsimile	LDF	UXC7	5815011877844
JR	L69306*	HHC	Line of Sight Multichannel Radio Terminal	RDO	TRC190V1	5820012470981
		HEF		TML	TRC190AV1	5820013102538
JR	L69374*	HHD	Line of Sight Multichannel Radio Terminal	RDO	TRC190V2	5820012470979
		HEL		TML	TRC190AV2	5820013094649
JR	L69442*	HHE	Line of Sight Multichannel Radio Terminal	RDO	TRC190V3	5820012470982
		HEH		TML	TRC190AV3	5820013102543
JR	L69510*	HHF	Line of Sight Multichannel Radio Terminal	RDO	TRC190V4	5820012470980
		HEM		TML	TRC190AV4	5820013094651
QE	L70538*	ZLH	Laundry Advanced System	LAU ADV SYS	LADS	3510014630114

*Denotes that items will be reported as systems.

TABLE 6.3K			List of Ground Equipment for DA Form 2406, continued			

ECC	LIN	EIC	Nomenclature	Abbreviation	Model No.	NSN
NG	L76321	EFC	Loader, Scoop, DED (CCE)	LDR SCP	175B	3805006025013
		EFS			H100CGPB	3805010529043
NG	L76556	EFW	Loader, Scoop, DSL 2.5 CU YD	LDR SCP	950BNS	3805011267915
		EFQ			MW24C	3805011504814
		EGG			950BNSCE	3805012605163
NG	L76693	EFV	Loader, Scoop, SEC 2.5 CU YD	LDR SCP	950BS	3805011267914
		EGF			950BSCE	3805012605162
DB	M02114	4SK	Mortar, 81MM	MORTAR	M252	1015011646651
JS	M04268*	HHJ	Management Facility	MGMT FAC	TSQ154	5895012470963
		HDY			TSQ154A	5895013301864
JM	M04941*	KE2	Meteorological Data System	MDS	TMQ31	6660011481772
JH	M21948*	L6E	Master Control Set	MCS	TSQ138	5895011657408
JX	M52582	HPR	Message Entry Device Variable Format	MSG ENT DV	GSC21	7010010176967
JX	M52650	HPW	Message Device Digital	MSG DV DIG	PSG2A	7025010443824
		HPZ			PSG2	7025010945473
		HP3			PSG2B	7025011269199
NB	M57048*	E46	Mix PLT ASPH ELEC 150	MIX PLT	KA60A	3895013692551
		EY6	Mix PLT ASPH DSL/ELEC	MIX PLT	KA60	3895009368613
DB	M67871	4SA	Mortar, 60MM on Mount	MRTR W/MT	M2	1010006732006
		4SB				1010006732010
DB	M67939	4SC	Mortar, 60MM: On Mount	MRTR W/MT	M224	1010010205626
DB	M68008	4SG	Mortar, 81MM: On Mount	MRTR W/MT	M29	1015008401836
		4SJ			M29A1	1015009997794
DB	M68282	4SH	Mortar, 4.2 Inch: On Mount	MRTR W/MT	M30WMT24A1	1015008401840
		4SD			M30WMT24	1015003229720
DB	M68405	4SL	Mortar	MRTR TWD	M120T	1015012261672
		4SE			M120C	1015012923801
JA	N04596	IPH	Night Vision Sight (Crew)	NT VIS ST	TVS5	5855006295327
NB	N75124	EXE	Paving Machine Bituminous Material, Dsl	PAVG MACH	IOWABSF400	3895010637891
		E47			780T	3895013791102
JC	P05439*	HHO	Operations Group	OPER GRP	OL412TTC46	5805012459059
		HED			OL412TC46A	5895013136195
		HEC			OL412TC46B	5805013266540
ST	P06082	YTY	Plate Process Sect Topo Reproduction Set STLR Mtd	P SECT TOPO	13225E3019	3610011051743
QC	P11866	FBD	Pneumatic Tool and Compressor Outfit: 250CFM	PN TL	250CFM	3820009508584

*Denotes that items will be reported as systems.

TABLE 6.3L			List of Ground Equipment for DA Form 2406, continued			
ECC	LIN	EIC	Nomenclature	Abbreviation	Model No.	NSN
OF	P19377	8CI	Operating and Treatment Unit, Dental	OPER UT	2100	6520013438126
QR	P21220*	YOA	Position and Azimuth Determining System	PADS	USQ70	6675010715552
QP	P27819	VCO	Power Plant, Electric, 30KW TM	PWR PLT EL	MJQ10A	6115003949582
QP	P27823	VEL	Power Plant, Electric, 60KW TM	PWR PLT EL	MJQ12A	6115002571602
QP	P28015	VJD	Power Plant, Electric, 10KW, TM	PWR PLT EL	MJQ18	6115000331398
QP	P28075	VLO	Power Plant, Electric	PWR PLT EL	MJQ15	6115004007591
QP	P28083	VKJ	Power Plant, Electric, 5KW, 60HZ, TM	PWR PLT EL	MJQ35	6115013134216
		VD5			MJQ35A	6115014149697
QP	P28151	VKI	Power Plant, Electric, 5KW, 60HZ, TM	PWR PLT EL	MJQ36	6115013134215
QP	P42126	VNA	Power Plant, Electric, 30KW, 50/60HZ, TM	PWR PLT EL	MJQ40	6115012996033
QP	P42194	VF2	Power Plant, Electric, 60KW, 50/60HZ, TM	PWR PLT EL	MJQ41	6115013037896
QP	P42262	VK2	Power Plant, Electric, 10KW, 60HZ, TM	PWR PLT EL	MJQ37	6115012996035
QP	P42330	VK3	Power Plant, Electric, 10KW, 400HZ, TM	PWR PLT EL	MJQ38	6115013134214
QP	P42614	VD2	Power Plant, Electric, TM	PWR PLT EL	MJQ39	6115012996034
QB	P44627	UAG	Power Unit, Auxil, Aviation (AGPU)	PWR UNT AX	MEP360A	1730011441897
QQ	P50154	YEP	Press Sect Topo, Repro Set, Semi-Trlr Mtd	P SECT TOPO	PSREPRO	3610003444705
		YF9		P SEC	PSREPRO	3610011051744
JY	P60206	QT3	Printer Station	PRINT STAT	OA9472TYQ	7010014204987
JC	P60408*	GEA	Node Center Switch	OPER GRP	413TTC47E	5805014544416
OD	P63884	8DF	Processing System, X-Ray Film	PRC RD FLM	3474B	6525008238144
JC	P70292*	HHP	Operations Group	OPER GRP	413TTC47	5805012444259
		HEB			413TTC47A	5895013094652
		HEA			413TTC47B	5895013246855
JS	P70360*	GAX	Operations Group	OPER GRP	413TTC47C	5895013301866
QD	P97051	ZCB	Pumping Assy Flambl Liq Eng Drvn	PMP FLAM L	A12BMVG4D	4320000698494
		ZCD			US37ACG	4320001954914
		ZCK			A12CMVG4D	4320006007590
		ZCM			A12MGDAD	4320006911071
		ZC4			ADC1500	4320010923551
		ZDR			LPPTM	4320012157671
		ZDT			LC350GPM	4320012595965
		ZDS			W8646	4320012464398
		ZTJ			LC35GPM	4320012595965
OD	P98514	8DL	Process Machine, Rad Film	PRC RD FLM	AFP14X3MIL	6525013036235
		8DM			MM190	6525014226122

*Denotes that items will be reported as systems.

TABLE 6.3M			List of Ground Equipment for DA Form 2406, continued			

ECC	LIN	EIC	Nomenclature	Abbreviation	Model No.	NSN
JP	Q16110	IAF	Radar Set	RDR ST	PPS5	5840001681567
		IAG			PPS5A	5840002389366
		IAM			PPS5B	5840010094939
JP	Q16173	IAP	Radar Set	RDR ST	PPS15AV1	5840010513067
JR	Q32756	GF2	Radio Set	RDO ST	GRC106	5820004022263
		GFZ			GRC106A	5820002237548
JR	Q38296	GGA	Radio Set	RDO ST	PRC74B	5820009350030
		GFX			PRC74C	5820001771641
		GAH			PRC77	5820009303724
LM	R10527	XMG	Ramp, Bay, Bridge Floating	RBBF	BF	5420004975276
JP	R14148*	IYA	Radar Set Mortar Locating	RDR ST	TPQ36V1	5840010434257
		IY2			TPQ36V3	5840011854244
		IYE			TPQ36V5	5840012291276
JP	R14216*	IT6	Radar Set	RDR ST	TPQ36V7	5840012291278
JP	R14284*	GGY	Radar Set	RDR ST	TPQ36V8	5840013900529
JR	R30895	GGD	Radio Set	RDO ST	GRC213	5820011283935
		GGR			GRC213AV1	5820012629548
JR	R30963	HBT	Radio Set	RDO ST	GRC224	5820012506254
JR	R33351*	HHG	Radio Access Unit	RDO ACC UT	TRC191	5820012475731
		HEG			TRC191AV1	5820013102542
		HEP			TRC191AV2	5820013260711
JH	R36854*	L5D	Receiving Set, Radio	RCV ST RDO	TRQ32	5820000678914
		L5F			TRQ32V1	5895011677655
JR	R38349	GGC	Radio Set	RDO ST	PRC70	5820010628246
JR	R38403	L2S	TAC SATCOM Radio Set	RDO ST	PSC3	5820011454943
JH	R38883*	KBC	Receiving Set	RCV ST RDO	TRQ37	5820011604684
JR	R39452*	HDK	Radio Terminal Set	RDO TML ST	TRC173	5820011619422
		HDS			TRC173A	5820013160890
		HE1			TRC173B	5820013874952
JR	R39520*	HDJ	Repeater Set Radio	RPT ST RDO	TRC174	5820011619420
		HDT			TRC174A	5820013160880
		HE2			TRC174B	5820013874520
JR	R39588*	HDL	Radio Terminal Set	RDO TML ST	TRC175	5820011619421
		HDU			TRC175A	5820013160891
		HE5			TRC175B	5820013876700
JU	R40028	KIR	REC SYS, SP PURPOSE	REC SYS	ANTSQ205	5895014077006
JU	R40255	GB7	RECEIVER, TRANSMITTER, RADIO	REC TRANS	RT476A ARC201AV	5821013064654

*Denotes that items will be reported as systems.

TABLE 6.3N			List of Ground Equipment for DA Form 2406, continued			

ECC	LIN	EIC	Nomenclature	Abbreviation	Model No.	NSN
QW	R41282*	551	Reconnaissance System	REC SYS	M93A1	6665013721303
QW	R41532	559	Reconnaissance System	REC SYS	M93	6665013232582
GF	R50544*	3LA	Recovery Vehicle, Full Tracked Light Armored	REC VEH LT	M578	2350004396242
GF	R50681*	AQA	Recovery Vehicle, Full Tracked Medium	REC VEH MD	M88A1	2350001226826
GF	R50885*	AQC	Recovery Vehicle, FT	REC VEH FT	M88A2	2350013904683
JR	R55200	GGF	Radio Set	RDO ST	PRC104A	5820011417953
		GGS			PRC104BV4	5820012629550
JR	R55268	L2A			PRC119	5820011519915
JC	R57843	L3B	TAC SATCOM Base	SAT TERM	VSC7	5820010905449
OJ	R61868	8AB	Refrigerator Mechanical	REF MECH	BR37SS1B01	4110011173902
		8AE			139875	4110011596922
		8AF			FT2TRBLB	4110013523653
OJ	R64126	8AD	Refrigerator Solid State Bio	REF SOL ST	ALL MODELS	4110012877111
JR	R78116*	HDM	Repeater Set, Radio	RPT ST RDO	TRC138A	5820011619419
		HDV			TRC138B	5820013160881
		HE3			TRC138C	5820013874544
JR	R83005	L2Q	Radio Set	RDO ST	PRC119A	5820012679482
JR	R83073	GC9	RADIO SET	RDO ST	PRC119D	5820014210801
JR	R92967*	HGX	Radio Terminal Set	RDO TML ST	TRC170V2	5820011483977
JR	R92996*	HCP	Radio Terminal Set	RDO TML ST	TRC145BV1	5820011044748
		HBG			TRC145V1	5820004515523
JR	R93035*	HGY	Radio Terminal Set	RDO TML ST	TRC170V3	5820011483976
HS	S10059	CVT	Trailer Tank Fuel 5000 GAL	TRL TNK FU	M967	2330010505632
		CVW			M967A1	2330011550046
NH	S11711	ET5	Roller Motorized, Steel wheel	RLS SP	C350B	3895005780372
		E5B			CB534B	3895013962822
NH	S11793	EUR	Roller Pneumatic, VP, Self-Propelled	RLR SP	C530A	3895010133630
NH	S12575	ETR	Roller Towed, Sheepsfoot	RLR TWD	111	3895001347981
		ET4			MDG96	3895008935006
		ETY			H2S	3895009679021
NH	S12916	EUP	Roller Vibratory Self-Propelled	RLR SP	RS28	3895010128875
		EUU			SP848	3895010752823
JS	S24750*	HD9	Switching Group	SWTCH GRP	305TTC46	5805012459053
		HEN			305TTC46A	5895013094654
		HD8			305TTC46B	5895013236459

*Denotes that items will be reported as systems.

TABLE 6.30			List of Ground Equipment for DA Form 2406, continued			

ECC	LIN	EIC	Nomenclature	Abbreviation	Model No.	NSN
JC	S24818*	HDX	Switching Group	SWTCH GRP	ON306TTC47	5895012459054
		HD6			306TTC47A	5895013094653
		HD7			306TTC47B	5895013240863
JC	S25379*	HHL	Small Extension Node Switch	SENS	TTC48V2	5805012459058
		HD4			TTC48AV2	5805013102539
		HD5			TTC48BV2	5805013240862
JC	S25447*	HHK	Small Extension Node Switch	SENS	TTC48V1	5805012444257
		HD2			TTC48AV1	5805013094650
		HD3			TTC48BV1	5805013240861
JC	S25515*	HO2	Small External Node	SMEXT	ANTTC48DV1	5805014543561
QS	S25681	2FQ	Shop Equip Contact Main	SHP EQ CM	No Model	4940013338470
NE	S29971	EHZ	Scraper, Tractor	SCPR	NONSECT	3805011442992
NE	S29971	EJL	Scraper, Tractor	SCPR	613BSNS	3805012674178
NE	S30039	EH2	Scraper, Elevating, SP, Sect	SCPR S	ECT	3805011448837
		EJK			613BSS	3805012674177
QS	S30914	2MB	Shop Equipment Contact Maint Eng, Truck Mounted	SHP EQ ENG	SEQENG	4940012098824
QS	S30982	2MC	Shop Equipment Contact Maint ORD, Truck Mounted	SHP EQ ORD	SEQORD	4940012098825
QS	S31232	2MA	Shop Equipment General Purpose, Truck Mounted	SHP EQ GP	SEQGP	2320012098823
JC	S34963*	L3E	Satellite Communication Terminal	SAT COM TM	TSC93BV1	5895012848306
		L3A			TSC93A	5895011135344
JC	S37228*	GAW	Switching Group	SWTCH GRP	306TTC47C	5895013294811
JS	S38172*	GAV	Small Extension Node Switch	SENS	TTC48CV4	5805013294808
OM	S39122	8EC	Sterilizer Surgical Dressing 16–36 in.	STR SUR DR	FX1636	6530009262151
JY	S44664*	HHQ	System Control Group Planning	CNTRL GRP	OL414TYQ35	5805012466817
JY	S44732*	HHS	System Control Group Management	CNTRL GRP	OL416TYQ35	5805012475730
JY	S44914*	HHR	System Control Group Technical	CNTRL GRP	OL415TYQ35	5805012444258
NC	S56246	EH3	Scraper Earth Moving SP	SCRPR SP	621B	3805011531854
HS	S70027	CVB	Semitrailer Flat Bed, 22½ Ton	STRLR FB	M871	2330001226779
		CWY			M871A1	2330012260701
		CVZ			M871A2	2330012943367
HS	S70159	CFE	Semitrailer Flat Bed, 34 Ton	STRLR FB	M872	2330010398095
		CFF			M872A1	2330011098006
		CFG			M872A2	2330011195837
		CFH			M872A3	2330011421385
HS	S70517	CFD	Semitrailer Low Bed, 25T	STRLR LB	M172A1	2330003176448

*Denotes that items will be reported as systems.

ECC	LIN	EIC	Nomenclature	Abbreviation	Model No.	NSN
TABLE 6.3P			**List of Ground Equipment for DA Form 2406, continued**			
HS	S70594	CFB	Semitrailer Long Bed 40 Ton	STRLR LB	M870	2330001331731
		CFC			M870A1	2330012249245
HS	S70661	CFA	Semitrailer Long Bed 60 Ton	STRLR LB	M747	2330000897265
HS	S70859	CXU	Semitrailer Low Bed, 70 Ton HET	STRLR LB	M1000	2330013038832
HS	S72024	CVA	Semitrailer Stake 12-Ton, 4 Wheel W/E	STRLR STK	M127A1	2330000487743
		CVD			M127A1C	2330007529750
		CVE			M127A2C	2330007886299
		CVF			M127	2330007979207
HS	S72846	CVL	Trailer Tank Fuel 5000 Gal	TLR TNK FU	M131A5	2330002266079
		CVN			M131A3C	2330005333380
		CVS			M131A4	2330009949459
HS	S72983	CVM	Trailer Tank Fuel 5000 Gal	TLR TNK FU	M131A5C	2330002266080
		CVR			M131A4C	2330009949458
HS	S73119	C4V	Semitrailer Tank, Petroleum 7500 Gal	STRLR TNK	M1062	2330012757475
HS	S73372	CVU	Trailer Tank Fuel 5000 Gal	TLR TNK FU	M969	2330010505634
		CVY			M969A1	2330011550048
		CW2			M969A2	2330013779337
JC	S78466*	L2Z	Satellite Communication Terminal	SAT COM TM	TSC85A	5895011135343
		L3F			TSC85BV1	5895012848305
JC	S78717*	GDX	Switching Group S	W GP	ON306TTC47E	5895014543549
HE	T05028	BEB	Truck Utility Tactical ¾T 1¼T	TRK UT TAC	M1009	2320011232665
HF	T05096	BBC	Truck Utility TOW Carrier	TRK UT	M966	2320011077153
		BBX			M966A1	2320013723932
KC	TO6859	ATC	Test Set Common Core (STE-M1/FVS)	TS COM COR	COMMONCORE	6625011354389
HF	T07543	BBK	Truck Utility S250 Shelter Carrier 4x4	TRK UT SHL	M1037	2320011467193
HF	T07679	BBM	Truck Utility Heavy Variant, 5T	TRK UT HV	M1097	2320013469317
		BBU			M1097A1	2320013719583
		BB6			M1097A2	2320013808604
QJ	T09094	ZHS	Tactical Water Distribution System	TWDES	MILT53023	4320011223547
		ZSG			TWDS10	4320012216006
		ZSH			TWDS20	4320013619232
QS	T10138*	2CU	Shop Equipment, Contact Maintenance, Truck Mounted	SP EQ MNT	993	4940001957712
		2CZ			ANC6217	4940004950118
		2CT			CMU3	4940001693042
		2CD			CMU5	4940001654019
		2CX			MILS45855	4940004950118
		2C5			SEMC1975	4940010162262

*Denotes that items will be reported as systems.

TABLE 6.3Q	List of Ground Equipment for DA Form 2406, continued

ECC	LIN	EIC	Nomenclature	Abbreviation	Model No.	NSN
QS	T10275*	2DA	Shop Equipment, Electronic Repair, Semitrailer Mounted	SP EQ ELEC	MILS52330	4940002949517
		2CE			SER1961	4940001654020
		2CB			SER1968	4940001598847
		2C6			SER1976	4940010225322
		2C8			SER197881	4940010964475
		2DL			SER1982	4940011503113
		2CM			FSVAN1959	4940001693036
		2CN			FSVAN15777	4940001693037
		2FP			CLB05	4940012342322
QS	T10412*	2CA	Shop Equipment, Electronic Repair, Semitrailer Mounted	SP EQ ELEC	SEER1968	4940001598846
		2CP			EER1963	4940001693038
		2C9			ELECREP	4940011107422
		2CY			MILS52377	4940002949542
QS	T10549*	2C2	Shop Equipment, General Purpose Repair, Semitrailer Mounted	SP EQ GP R	MED1952	4940004976412
		2CJ			ENG4359	4940001654024
		2CV			MILS45538	4940002874894
		2C4			SGPRSMD	4940010063229
		2C3			SGPRSM61	4940004976413
		2CF			SGPRSM68	4940001654021
QS	T13152*	2CG	Shop Equipment, Organizational Repair, Light Truck Mounted	SP EQ ORG R	ENG40	4940001654022
		2CR			MEDL1954	4940001693040
		2CH			MEDL1956	4940001654023
		2CS			SEORL66	4940001693041
		2C7			SEORL118	4940010282672
		2CC			SOUTHWEST	4940001642719
		2CQ			SMGPR61	4940001693039
		2CW			MILS45537	4940002949516
		2FN			SEORTM	4940012360166
FB	T13168*	AAB	Tank, Combat, Full Tracked	TNK CBT FT	M1A1	2350010871095
FB	T13169*	ABL	Tank Combat Full Tracked 105-MM TTS	TNK CBT FT	M60A3TTS	2350010612306
FB	T13305*	AAF	Tank Combat Full Tracked 120-MM	TNK CBT FT	M1A2	2350013285964
FB	T13374*	AAA	Tank Combat Full Tracked 105-MM M1	TNK CBT FT	M1	2350010612445
		AAC			M1IP	2350011368738
JY	T13413	HYE	Tactical Computer Processor	TCP	UYQ43V1	5895012119821
JY	T13481	HQL	Tactical Computer Processor	TCP	UYQ43V2	5895012468276
AS	T19416	LGV	Transmitting Set Radio	TRMT ST	ANFRN41V2	5825010705842
JH	T22676	IXM	Transponder Set	TRNSP ST	PPN19	5895011951199
		IWM			PPN19V1	5895012086159

*Denotes that items will be reported as systems.

TABLE 6.3R	List of Ground Equipment for DA Form 2406, continued

ECC	LIN	EIC	Nomenclature	Abbreviation	Model No.	NSN
ND	T33786	EED	Tractor Wheeled, W/Forklift and Crane	TRAC WHLD	HMMH	2420012058636
ND	T34437	EDL	Tractor Wheeled	TRAC WHLD	FLU419	2420011602754
HF	T38660	BEA	Truck Ambulance Tactical	TRK AMB	M1010	2310011232666
HF	T38707	BBB	Truck Ambulance 2 Litter	ARMD TRK AMB	M996	2310011112275
		BB2			M996A1	2310013723935
HF	T38844	BBA	Truck Ambulance 4 Litter	TRK AMB	M997	2310011112274
		BBZ			M997A1	2310013723934
		BB8			M997A2	2310013808225
HL	T39518	B2D	Truck Cargo Tactical W/W	TRK CGO	M977WW	2320010970260
HL	T39586	B2J	Truck Cargo Tactical	TRK CGO	M985	2320011007673
HL	T39654	B2E	Truck Cargo Tactical W/W	TRK CGO	M985WW	2320010970261
HH	T40329	BHG	Truck Van, LMTV, 2½ Ton W/W	TRK VAN	M1079WW	2320013601891
HM	T40999	B4H	Truck Cargo Heavy PLS, Transporter, 16.5T	TRK CGO	M1075	2320013042278
HI	T41036	BR9	Truck Cargo, MTV, 5T	TRK CGO	M1093	2320013553063
HM	T41067	B4G	Truck Heavy PLS Transporter, 16.5T	TRK CGO	M1074	2320013042277
HI	T41104	BT4	Truck, Cargo, MTV, 5T, W/W	TRK CGO	M1093WW	2320013601896
HI	T41135	BT3	Truck, Cargo, MTV, 5T, W/W	TRK CGO	M1083WW	2320013601895
HI	T41203	BR3	Truck, Cargo, MTV, 5T, W/MHE	TRK CGO	M1084	2320013543387
HG	T41995	BHF	Truck, Cargo, LMTV, 2½ T	TRK CGO	M1081	2320013553064
HG	T42063	BHJ	Truck, Cargo, LMTV, 2½ T	TRK CGO	M1081WW	2320013601899
PG	T48941	DJN	Truck, Lift, Fork, DED 50,000 LB Rough Terrain CONT HDLR	TRK LF	DV43	3930010823758
PG	T48944	DJW	Truck, Lift, Fork DED 6,000 LB Variable Reach RT Ammo Hdlg	TRK LF	RTFL	3930011580849
PC	T49096	DXG	Truck, Lift, Fork, DSL, 6,000 LB	TRK LF	CBDFL	3930011727892
PG	T49119	DJU	Truck, Lift, Fork, 10,000 LB RT	TRK LF	M10A	3930010543833
PG	T49255	DJV	Truck, Lift, Fork, 4,000 LB RT	TRK LF	M4K	3930010764237
		DJ5			MHE271	3930013308906
		DJ6			MHE270	3930013308907
GZ	T52849	4WQ	Test Set Electronics Systems, Direct Support	DSESTS	DSESTS	6625011200764
HM	T53858	BHA	Truck Maintenance Telephone, Utility	TRK UT	M876	2320000000114
JR	T55957	HHM	Terminal Radio Telephone, Mobile Subscriber	TML RDO TL	VRC97	5820012466818
HL	T58161	B2C	Truck Tank, Fuel Service	TRK TNK FU	M978WW	2320010970249
HM	T59048	B5C	Truck Tractor Cargo Tactical HET	TRK TRAC	M1070	2320013189902
HL	T59278	B2G	Truck Cargo Tactical	TRK CGO	M977	2320010996426
HF	T59346	BEC	Truck Cargo Tactical	TRK CGO	M1008A1	2320011232671

TABLE 6.3S			List of Ground Equipment for DA Form 2406, continued			
ECC	**LIN**	**EIC**	**Nomenclature**	**Abbreviation**	**Model No.**	**NSN**
HF	T59414	BEE	Truck Cargo Tactical Shelter W/E 1.25T	TRK CGO	M1028	2320011275077
HF	T59482	BED	Truck Cargo Tactical W/E 1.25T	TRK CGO	M1008	2320011236827
HF	T59550	BEF	Truck Cargo, 5/4T	TRK CGO	M1028A1	2320011580820
HG	T60081	BHD	Truck Cargo, LMTV, 2	TRK CGO	M1078	2320013543385
HG	T60149	BHH	Truck Cargo, LMTV, 2½T	TRK CGO	M1078WW	2320013601898
HM	T61035	B5B	Truck Tractor (HET)	TRK TRAC	M911	2320010253733
HM	T61103	B4A	Truck Tractor, Line Haul	TRK TRAC	M915	2320010284395
		B4B			M915A1	2320011252640
		B4E			M915A2	2320012725029
HM	T61171	B4D	Truck Tractor (MET)	TRK TRAC	M920	2320010284397
HJ	T61239	BTJ	Truck Tractor, MTV, 5T	TRK TRAC	M1088	2320013554332
HJ	T61307	BTY	Truck Tractor, MTV, 5T, W/W	TRK TRAC	M1088WW	2320013601892
HF	T61494	BBD	Truck Utility Cargo Troop Carrier W/E	TRK UT	M998	2320011077155
		BBN			M998A1	2320013719577
HF	T61562	BBE	Truck Utility Cargo Troop Carrier W/W 1.25T	TRK UT WW	M1038WW	2320011077156
		BBP			M1038A1WW	2320013719578
HF	T61630	B6B	Truck Utility	TRK UT	M1113	2320014120143
HI	T61704	BR7	Truck Cargo, MTV, LWB 5T	TRK CGO	M1085	2320013544530
HI	T61772	BT5	Truck Cargo, MTV, LWB 5TWW	TRK CGO	M1085WW	2320013601897
HI	T61840	BR8	Truck Cargo, MTV, LWB, W/MHE, 5T, W/W	TRK CGO	M1086WW	2320013544531
HI	T61908	BR2	Truck Cargo, MTV, 5T	TRK CGO	M1083	2320013543386
HL	T63093	B2B	Truck Wrecker W/W	TRK WRK WW	M984WW	2320010970248
		B2L			M984A1WW	2320011957641
HJ	T64911	BR5	Truck, Dump, MTV, 5T	TRK DMP	M1090	2320013544529
HJ	T64979	BTZ	Truck, Dump, MTV, 5T, W/W	TRK DMP	M1090WW	2320013601893
HJ	T65526	BTK	Truck, Dump, MTV, 5T	TRK DMP	M1094	2320013553062
HJ	T65594	BT2	Truck, Dump, MTV, 5T, W/W	TRK DMP	M1094WW	2320013601894
LE	T68330	WGE	Tug, Large Diesel	TUG	No Model1	1925012477110
PG	T73347	DJ8	Trk Lift Fork Rt	TRK LF	10000M	3930014172886
PC	T73645	DXA	Truck, Lift, Fork 4,000 LB, Clean Burn Diesel	TRK LF	CBD4000	3930011727891
HL	T87243	B2H	Truck Tank Fuel Servicing	TRK TNK FU	M978	2320011007672
HL	T88677	B2A	Truck Tractor Tactical W/W	TRK TRAC	M983WW	2320010970247
HL	T91308	DV4	Trk Common Bridge Trans	TRK CARGO	M1977WW	2320014438023
		DVZ			M1977WOW	2320014421940
HM	T91656	B4C	Truck Tractor (LET), 6X6	TRK TRAC	M916	2320010284396
		B4F			M916A1	2320012725028
		B4J			M916A2	2320014311163

TABLE 6.3T			List of Ground Equipment for DA Form 2406, continued			

ECC	LIN	EIC	Nomenclature	Abbreviation	Model No.	NSN
HF	T92242	BBF	Truck Utility ARMT Carrier ARMD	TRK UT	M1025	2320011289551
		BBV			M1025A1	2320013719584
		BB3			M1025A2	2320013808233
HF	T92310	BBG	Truck Utility ARMT Carrier ARMD	TRK UT WW	M1026WW	2320011289552
		BBQ			M1026A1WW	2320013719579
HF	T92446	B6C	Trk Utl Arm HV	TRK UT HV	M1114	2320014133739
HH	T93484	BHE	Truck, Van, LMTV, 2½T	TRK VAN	M1079	2320013543384
HT	T93761	C9C	Trailer Palletized Loading	TRLR PLS	M1076	2330013035197
NJ	T94171	ZJM	Truck Well Drilling Support	TRK DR SPT	WDS	3820011784980
HJ	T94709	BR4	Truck Wrecker, MTV, 5T	TRK WKR	M1089	2320013544528
PL	U12203	DSH	Spreader Lifting Frt Container	SPDR LFT	SLFCTL	3990002969398
		DSL			SLFCTLSA	3990011280089
		DSP			ISO214A	3990012582010
LF	V00426	WAX	Vessel Logistic Support, 245 to 300 FT LG, 3,000 to 5,500 Ton Cap	LSV	LSVNDI	1915011538801
QH	V12141*	ZAC	Tank and Pump Unit	TNK PMP UT	MDL1800	4930000701181
		ZAE			MD2938	4930000784939
		ZAO			MD1151	4930005422800
		ZBG			ENG2519	4930009878576
		ZAR			HLND2000	4930008778678
		ZBE			ORRBL100	4930009263692
		ZAD			BOW36W50	4930000784938
		ZBD			ALTECH	4930009263581
		ZAL			13217E7100	4930004269960
		ZBH			13217E7130	4930011307281
		ZA5			126ETP	4930012740021
FB	V13101*	ABB	Tank, Combat, Full Tracked 105MM	TNK CBT FT	M60A3	2350001486548
JC	V57504*	HJM	Terminal Telegraph	TML TG	TSC58	5805000105287
		HLV			TSC58A	5805010956232
OK	V99288	8BM	Ventilator Mobile Volume	VENT ANES	V5A	6515011167903
OK	V99538	8BO	Ventilator Volume Portable	VENT VOL	750M 6	530013270686
		8BP			15304	6530013748903
QJ	W35417*	ZIP	Water Purif Equip Set: Reverse Osmosis 600 GPH	WTR PURIF	ROWPU600	4610010268980
		ZTY			WSPES1	4610012952720
		ZU4			WPES10	4610013416289
QJ	W37311	ZIJ	Water Storage/Distribution Set	WTR S/D ST	CPL81045	4610011141450
		ZU5			WSDS810	4610013601581
		ZU8			800KWSDS	4610013823547

*Denotes that items will be reported as systems.

TABLE 6.3U			List of Ground Equipment for DA Form 2406, continued			

ECC	LIN	EIC	Nomenclature	Abbreviation	Model No.	NSN
QJ	W47225*	ZHN	Water Purif Reverse Osmosis 3000GPH, TM	WTR PURIF	ROWPU3000	4610012198707
		ZH2			ROWPU1	4610013711790
QJ	W55968	ZIK	Water Storage/Distribution Set	WTR SD ST	40000GPD	4610011141451
ND	W76268	EBB	Tractor FL, TRKD Low SPD DSL	TRAC FL	D5BS	2410011276512
		EBS			D5BS1	2410012701192
ND	W76285	EA8	Tractor Full Tracked, Low Speed	TRAC FT	1150ROPS	2410010244065
		EBA			D5BNS	2410011267902
		EBT			D5BNS1	2410012968479
ND	W76336	EBC	Tractor Full Tracked, Low Speed, DSL	TRAC FT	550C	2410011399859
		EBU			450	2410014120930
GJ	W76473	ASA	Tractor, Full Tracked, High Speed Armored, Dozer/Scraper Combination Winch	TRAC FT	M9	2350008087100
ND	W76816	EA7	Tractor, Full Tracked, Low Speed W/Bulldozer, W/Winch	TRAC FT	D7FWNTRZD	2410003006664
		EA6			D7FWR	2410001859792
		EA2			D7FDV29	2410001777284
		EBM			D7G	2410012237261
		EBY			D7HWCAB	2410014230931
		EBV			D7GWW	2410012532117
ND	W83529	EAW	Tractor, Full Tracked, Low Speed, W/Bulldozer, W/Ripper	TRAC FT	D7FWR	2410001859794
		EAU			D7FDV29	2410001777283
		EAZ			D7GWROPS	2410012230350
		EB2			D7R	2410014514048
		EBX			D7HRCAB	2410014230930
		EBW			D7	2410012532118
ND	W88575	EAC	Tractor, Full Tracked, Low Speed, W/angle Dozer, W/Winch (CCE)	TRAC FT	D8K8A58	2410005747597
ND	W88699	EAD	Tractor, Full Tracked, Low Speed, W/bulldozer, W/Ripper (CCE)	TRAC FT	D8K8S8	2410005747598
ND	W91074	EDH	Tractor, Wheeled W/Backhoe, W/Loader, W/Hydraulic Tool Attachment (CCE)	TRAC WHL	JD410	2420005670135
HT	W95537	CDA	Trailer Cargo $3/4$T	TLR CGO	M101	2330007389509
		CDC			M101A1	2330008986779
		CDB			M101A2	2330011024697
LM	X23277	XMA	Transporter, Bridge Floating	TRSP BRDG	PACAR9999	5420000715321
		XMM			SWRBT	5420011756524
OQ	X37050	8DA	X-Ray Apparatus Field Dental	XRY AP DTL	D3152	6525010992320
		8DE			G336	6525012070824
		8DJ			ALPHAPM	6525013707552

*Denotes that items will be reported as systems.

TABLE 6.3V			List of Ground Equipment for DA Form 2406, continued			

ECC	LIN	EIC	Nomenclature	Abbreviation	Model No.	NSN
HG	X40009	BMA	Truck, Cargo, 2½ Ton	TRK CGO	M35A2	2320000771616
		BHK			M35A3	2320013832047
HG	X40077	BMR	Truck, Cargo, Drop Side 2½T	TRK CGO	M35A2C	2320009260873
		BHP			M35A3C	2320013832050
HG	X40146	BMB	Truck, Cargo, 2½T W/W 6x6	TRK CGO WW	M35A2WW	2320000771617
		BHL			M35A3WW	2320013833850
HG	X40214	BMS	Truck, Cargo, Drop Side 2½T W/W	TRK CGO WW	M35A2CWW	2320009260875
		BHQ			M35A3CWW	2320013832049
HG	X40283	BMC	Truck, Cargo, 2½T XLWB	TRK CGO	M36A2	2320000771618
		BHM			M36A3	2320013832048
HG	X40420	BMD	Truck, Cargo, 2½T XLWB W/W	TRK CGO WW	M36A2WW	2320000771619
		BHN			M36A3WW	2320013832046
HI	X40794	BQL	Truck, Cargo, Drop Side, 5 Ton WE 6x6	TRK CGO	M54A2C	2320007612854
		BSD			M813A1	2320000508913
		BRY			M923	2320010502084
		BSS			M923A1	2320012064087
		BS7			M923A2	2320012300307
HI	X40831	BQH	Truck, Cargo, 5T, LWB WE 6x6	TRK CGO	M54A2	2320000559266
		BSB			M813	2320000508902
		BRX			M924	2320010478773
		BSU			M924A1	2320012052692
HI	X40931	BQS	Truck, Cargo, Drop Side, 5 Ton W/W 6x6	TRK CGO WW	M54A2CWW	2320009260874
		BSC			M813A1WW	2320000508905
		BRT			M925WW	2320010478769
		BST			M925A1WW	2320012064088
		BS8			M925A2WW	2320012300308
HI	X40968	BQG	Truck, Cargo, 5T LWB W/W	TRK CGO WW	M54A2WW	2320000559265
		BSA			M813WW	2320000508890
		BRW			M926WW	2320010478772
		BSV			M926A1WW	2320012052693
HI	X41105	BSK	Truck, Cargo, 5T XLWB	TRK CGO	M814	2320000508988
		BRV			M927	2320010478771
		BSW			M927A1	2320012064089
		BS9			M927A2	2320012300309
HI	X41242	BQB	Truck, Cargo, 5T XLWB, W/W	TRK CGO	M55A2WW	2320000559259
		BSJ			M814WW	2320000508987
		BRU			M928WW	2320010478770
		BSX			M928A1WW	2320012064090
		BTM			M928A2WW	2320012300310

TABLE 6.3W	List of Ground Equipment for DA Form 2406, continued				

ECC	LIN	EIC	Nomenclature	Abbreviation	Model No.	NSN
HJ	X43708	BQE	Truck, Dump, 5 Ton	TRK DMP	M51A2	2320000559262
		BSF			M817	2320000508970
		BTH			M929	2320010478756
		BSY			M929A1	2320012064079
		BTN			M929A2	2320012300305
HJ	X43845	BQF	Truck, Dump 5T WW	TRK DMP WW	M51A2WW	2320000559263
		BSR			M817WW	2320000510589
		BTG			M930WW	2320010478755
		BSZ			M930A1WW	2320012064080
		BT7			M930A2WW	2320012300306
NN	X44403	EZY	Truck, Dump, 20 Ton (CCE)	TRK DMP	F5070	3805001927249
		EZZ			M917	3805010284389
NN		E5C	Trk Dump 20T (CCE)	TRK DUMP WW	M917A1	3805014311165
		E5D			M917A1MCS	3805014328249
PG	X48914	DJC	Truck, Lift Fork, Dsl Drvn, 6000 LB	TRK LF	ARTFT6	3930004195744
		DJS			ARTFT6ROPS	3930010543830
		DJJ			MLT6	3930009030900
		DJB			MLT62	3930003271575
		DJL			MLT6CH	3930009370220
		DJQ			MLT6CHROPS	3930010534823
		DJT			MLT6ROPS	3930010543831
		DJK			MLT6W	3930009263835
PB	X50489	DBE	Truck, Lift Fork, Elec, 4,000 LB	TRK LF	040M02	3930000645871
		DBG			337450	3930000866677
		DAC			FTD040EE	3930002366253
		DBN			4024	3930002668966
		DBS			FTHEG	3930002729972
		DBY			BF40	3930002738229
		DAE			CE40AEE180	3930003271600
		DAJ			FL40EE6250	3930004035662
		DA3			FTHYG	3930005541985
		DAM			FTD040	3930007096341
		DDC			BAK04EE	3930007096358
		DCB			CF40	3930009376176
		DDD			E40EV36V	3930012238437
PB	X50900	DAK	Truck, Lift Fork, Elec, 6,000 LB	TRK LF	FE6024	3930004798769
		DDA			EE5600	3930009357867
		DDB			60HEV36VEE	3930012238436

TABLE 6.3X			List of Ground Equipment for DA Form 2406, continued			

ECC	LIN	EIC	Nomenclature	Abbreviation	Model No.	NSN
HJ	X56586	BSP	Truck, Stake, 5 Ton W/W	TRK STK	M821WW	2320000509015
HJ	X59326	BQC	Truck, Tractor, 5 Ton WE	TRK TRAC	M52A2	2320000559260
		BSH			M818	2320000508984
		BTE			M931	2320010478753
		BS2			M931A1	2320012064077
		BTP			M931A2	2320012300302
HJ	X59463	BQD	Truck, Tractor, 5 Ton W/W	TRK TRAC	M52A2WW	2320000559261
		BSG			M818WW	2320000508978
		BTD			M932WW	2320010478752
		BS3			M932A1WW	2320012052684
		BTQ			M932A2WW	2320012300303
HJ	X62237	BSM	Truck, Van Expansible	TRK VAN	M820	2320000509006
		BTB			M934	2320010478750
		BS4			M934A1	2320012052682
		BTR			M934A2	2320012300300
HJ	X62271	BSN	Truck, Van, Expansible 5T W/Hydraulic Lift Gate	TRK VAN	M820A2	2320000509010
		BTC			M935	2320010478751
		BS5			M935A1	2320012052683
		BTS			M935A2	2320012300301
HH	X62340	BMJ	Truck, Van, Shop, 2½ Ton	TRK VAN	M109A3	2320000771636
HH	X62477	BMK	Truck, Van, Shop, 2½ Ton	TRK VAN	M109A3WW	2320000771637
HJ	X63299	BQA	Truck, Wrecker, 5 Ton W/W	TRK WRK	M543A2WW	2320000559258
		BSQ			M816WW	2320000510489
		BTF			M936WW	2320010478754
		BS6			M936A1WW	2320012064078
		BTT			M936A2WW	2320012300304
LE	X71046	WAQ	Tug	TUG	DSN377A	1925002161845
		WAM	Tug, Ocean Diesel	TUG	DSN3006	1925003753003
OQ	X90968	8DH	X-Ray Apparatus Med Capacity Port	XRY MED CAP	1200	6525013253740
		8DB			50MA 90KVP	6525012005800
OQ	X92158	8DG	X-Ray Apparatus Radiographic and Fluoroscopic	XRY RF	C58952	6525013126411
OQ	X92545	8DI	X-Ray Apparatus Radiographic Medical	XRY RM	LCROKS	6525013849296
QJ	Y35486*	ZIB	Water Purification Equipment Set: Truck Mounted 1,500 GPH	WPE 1500	1500GPH	4610002026925
QJ	Y36034*	ZIC	Water Purification Equipment Set: Truck Mounted 3,000 GPH	WPE 3000	3000GPH	4610002028701

*Denotes that items will be reported as systems.

TABLE 6.4A	List of Ground Subsystems for DA Form 2406

LIN	Noun Abbreviation	Subsystem	EOS Codes
A27624	ATC CEN	TSW7A	C
		Truck, 2½T, M35A2 (X40009)	M
		Generator Set PU405 (J35492), PU802 (G53788)	P
A41666	RDR SET	TPQ37V1, V2, V3, V4, V5, V6	C
		2 Radio Sets, ANVRC46 (Q53001)	C
		Generator Set, MEP115A (J38506), MEP816A (G18052)	P
		1 Truck 5T, M813A1/M813A1WW (X40794/X4079/X40931)	M
		1 Truck 2½T M35A2 (X40009)	M
A41666	RDR SET	TPQ37V8	C
		Generator Set, MEP115A (J38506) or MEP 816A (G18052 1)	P
		Truck M1097 (T07679)	M
		1 Truck Any Model (X40931) (X40794)	M
A48430	ALARM BIO AG	Air Conditioner (A24463)	E
		Generator Set (G78374)	P
		Truck, M1097A1 (T07679)	M
A48498	ALARM BIO AG	Air Conditioner, 18000 BTU (A24463)	E
		Generator Set, Diesel Engine (G78374)	P
		Truck Utility, Heavy HMMWV (T07679)	M
A93125	ARAAV	M551A1	M
		Main Gun	S
		Machine Gun, 7.62mm (L92352)	S
		Machine Gun, 50cal (L91975)	S
		Radio Set (Q53001, Q34308)	C
B31098	BRDG AVLB	Launch M60 Series Tank (L43664)	M
C00384	CARR AIR D	M60DS	M
		Navigation Set (N95862)	C
		Interrogator Set (J98501)	C
		M242 Gun (G96797)	S
C10990	CARR MTR	M1064, M1064A3 Radio Set, ANVRC46 (Q53001), ANVRC87A (R67160), 88A (R67194), 89A (R44863)	M
		90A (67908), 91A (R68010), or 92A (R45407)	C
		Intercom Set (K93373)	C
		KY57 (S01373)	K
		Machine Gun, .50 Cal (L91975)	S

TABLE 6.4B		List of Ground Subsystems for DA Form 2406, continued	
LIN	**Noun Abbreviation**	**Subsystem**	**EOS Codes**
C12815	CARR SM GE	M1059, M1059A3	M
		Machine Gun (L91975) Radio Set (Q34308, R44659, R45339, R67228, R67262, R44931, R67976, R68078) SMK GEN Set, M157 W/120G Tank[1]	D
C17936	FD ART COM ST	Generator Set, 60KW (G78306)	P
		Radio Set, AN/GRC-193A (H35404)	C
		Truck Utility, HMWWV (T07679), or Carrier (C11158)	M
		Radio Set, AN/VRC-90A, 91A, 88A, 92A (R67909, R68010, R67194, and R45407)	C
		Radio Set, ANPRC-104A (R55200)	C
C18072	FD ART COM ST	Generator Set 60KW (G78306)	P
		Radio Set AN/VRC92A, AN/VRC90A (R67908) (R45407)	C
C18234	CARR PER	M113A3	M
		Machine Gun, .50 Cal (L91975)	S
		Radio Set (Q34308, Q53001)	C
C27007	FD ART COM ST	Carrier Command Post, LT TRK	M
		Radio Set AN/VSQ-2(V)3 (E12117)	C
		Generator Set, PU-798 (G42238)	P
		Generator Set, Diesel, 60KW (G35851)	P
		Radio Set, AN/GRC-193A (H35404)	C
		Radio Set, AN-VRC-90A, 92A (H67908) (R45407)	C
C28728	CEN COMM	TSQ190V4	C
		Truck, M1113 (T61630)	M
		Trailer, M1102 (T95924)	B
		Speech Security Equipment KY68 (S64488)	K
		Truck Encryption Device, KG94 (T64771)	K
C30675	CTRMSR	TLQ17AV3	C
		2 Trucks, M1037 (TO7543) M1097 (TO7679)	M
C35900	COMM CTL	TSQ183, TSQ183B	C
		Generator Set, PU797 (G42238)	P
		Truck M1097 (T07679)	M
C36104	COMM CTL	TSQ184B, TSQ184E	C
		Generator Set, 4.2KW 28V (J46589)	P
		Carrier, M1068/A3 (C11158)	M

1. Only 1 M54A2 smoke generator is required for system to be FMC.

| TABLE 6.4C | List of Ground Subsystems for DA Form 2406, continued |

LIN	Noun Abbreviation	Subsystem	EOS Codes
C41061	CEN MSG SA	TYC39A, TYC39V1 TYC39V5	K
		2 Generator Sets, PU650 (J35629), PU805 (G78306) 2 Trucks, 5T,	C
		M923 (X40794)	P
		2 Trucks, 2½T, M35A2 (X40009)	M
		4 Air Conditioners, 18KBTU (A24463) KG 94, 82, 83, or 84	M
		(T64771), E02378, (E03568, S64488)	E
		One shelter version of this system, 1 5T truck and 2 air conditioners[2]	K
C41311	COTA	TTC39AV1, TTC39D, TTC39EV1	C
		Power Plant, MJQ10A (P27819) MJQ40 (P42126)	P
		1 Truck, M923 (X40794)	M
		2 Trucks, M35A2 (X40009)	M
		2 Air Conditioners, 18KBTU (A24463)	E
		KY57 (S01373), KY68 (S64488), KY82 (E02378), KY83 (E03568), KG94 (T64771)	K
C59125	COMM SYS	TSQ198	C
		Truck (T61494)	M
		Radio (VRC91 A or D)[3]	C
		AN/PSN-11 (N95862)	
		KY57 (S01373)	
		(Spare generator not included)[4]	
C76335	CFV	M3	M
		Main Gun 25MM, M242 (G96797)	S
		Radio Set (Q53001, Q56783)	C
		Missile F	
C78793	COTA	TTC41V2	C
		Truck, 1¼T, M1028/M1037, (T59414/T07543)	M
		Power Unit, PU620 (J47617)	P
		Air Conditioner 6KBTU (A23667)	E
C78861	COTA	TTC41V3	C
		Truck, 1¼T, M1028/M1037, (T59414/T07543), M1097 (TO7679)	M
		Power Unit, PU620 (J47617)	P
		Air Conditioner, 6KBTU (A23667)	E

2. For one shelter versions of this system, 1 5T truck and 2 air conditioners are required.

3. Sys has two VRC 91s but is FMC as long as one is operable.

4. The spare generator is not included here because it will not impair system readiness since the primary power source is the HMMVM and can also operate on commercial power.

original-2. Power sources air conditioners, or vehicles may be replaced by authorized substitutes listed in SB 700-20, appendix H.

original-3. Consult the respective technical manual for COMSEC quantities required.

TABLE 6.4D		List of Ground Subsystems for DA Form 2406, continued	
LIN	**Noun Abbreviation**	**Subsystem**	**EOS Codes**
C78929	COTA	TTC41V4	C
		Truck, 1¼, M885 (X39441)	M
		Power Unit, PU620 (J47617)	P
		Air Conditioner, 6 KBTU (A23667)	E
C84541	REF CONT	SC200, SC210	E
		Generator Set (J35825)	P
		Truck tractor 5T (X59326)	M
		Semi trailer flatbed (S70027)	B
C89935	CEN COMM	TSQ190V3	C
		2 Trucks M1113 (Z62562,T61630)	M
		Power Plants 10KW, PU798 (G42170)	P
C90003	CEN COMM	TSQ190V1	C
		2 Trucks M1113 (T61630)	M
		Power Plants 10KW, PU798 (G42170)	P
C90071	CEN COMM	TSQ190V2	C
		2 Trucks M1097 (T07679)	M
		Generator Set, l0KW, PU798 (G42170)	P
C90531	COMM CTL	TSQ182,TSQ182A	C
		Power Unit (G42170)	P
		Truck (T07679)	M
C90599	COMM CTL	TSQ183A,TSQ183C	P
		Power Unit (G42170)	M
		Truck (T07679)	
C90667	COMM CTL	TSQ184,TSQ184C	C
		Generator (G42238)	P
		Truck (T07679)	M
C90735	COMM CTL	TSQ184A, TSQ184D	C
		Generator Set (G42170)	P
		Truck (T07679)	M
D10281	DTSS LIGHT	D10281, ANTYQ-67V1	C
		Generator, 10KW (G42170)	P
		Truck (T61630)	M
		Air Conditioner, 18000 BTU (A24463)	E
D10741	CARR MRTR	M106A2	M
		Mortar (M68282)	S
		Radio Set (Q53001)	C

TABLE 6.4E	List of Ground Subsystems for DA Form 2406, continued

LIN	Noun Abbreviation	Subsystem	EOS Codes
D11248	DTSS HEAVY	Air Conditioner, 18000BTU (A24463)	E
		Generator Set, Diesel engine PU-802 (G53778)	P
		ANTYQ-48A (D11248)	C
D11538	CARR CP	M577A2 M577A3	M
		2 Radios (Q53001, Q56783) R44795, R44863, R44931, R44795, R67228, R67262, R44931, R67976, R68078, R45475	C
		Generator Set (J46589)	P
D12087	CARR PER	M113A2	M
		Machine Gun, 50 CAL (L91975)	S
		Radio Set (Q34308, Q56783) R44659, R45339, R45407, R67228, R67262, R44931, R67976, R68078, R45475	C
D78075	DP SYS	MYQ4	A
		Power Plant, MJQ10A (P27819), MJQ40 (P42126)	P
		Truck, Trac, 5T, M818 or M818WW (X59326/X59463)	M
		2 Air Conditioners, 18 KBTU (A24455)	E
D78325	DP SYS	MYQ4A	A
		Power Plant, MJQ12A (P27823), MJQ41 (P42194)	P
		Truck, Trac, 5T, M818 or M818WW (X59326/X59463)	M
		Truck, Van Exp, 5T, M934 (X62237)	M
		2 Air Conditioners, 18 KBTU (A24455)	E
D82404	DECON APP	AE32U8, M17, M17A1, A2, A3	D
		Truck (X40146) (T07543) (T61494, T61562, T07679)	M
		Trailer (W95537)	B
E56578	CBT EN VEH	M728	M
		Radio Set (Q53001, Q54174)	C
		Machine Gun, 7.62MM (L92352)	S
		Machine Gun, .50 CAL (L92112)	S
F40307	IFV	M2A1	M
		Main Gun 25MM M242 (G96797)	S
		Radio Set (Q53001, Q56783)	C
		Missile	F
F40375	IFV	M2A2, M242WODS	M
		Main Gun 25mm M242 (G96797)	S
		Radio Set (Q53001, Q56783)	C
		Missile	F
F43336	FES	TTC50	C
		Truck (T07679)	M
		Generator Set (G40744)	P

TABLE 6.4F	List of Ground Subsystems for DA Form 2406, continued

LIN	Noun Abbreviation	Subsystem	EOS Codes
F55539	FIRE CTL FA	ANGYK37V1	A
		Radio Set (R45407, R68010, R67194, R83005, R67908)	C
		Truck or Carrier (T61494, T07679, X40831, C11158, X62237, C11280, C12155, D11538)	M
F55607	FIRE CTL FA	ANGYK37V2	A
		Radio Set (R45407, R68010, R67194, R83005, R67908)	C
		Truck or Carrier (T61494, T07679, D11538, C12155, C11280, C11158) (T61494, T07679, D11538, C12155, C11280, C11158)	M
F60462	CFV	M3A1	M
		Main Gun 25MM M242 (G96797)	S
		Radio Set (Q53001,Q56783)	C
		Missile Launcher Assy Tow	F
F60530	CFV	M3A2, M3A2WODS	M
		Main Gun 25 MM M242 (G96797)	S
		Radio Set (Q53001, Q56783)	C
		Missile Launcher Assy Tow	F
F60564	IFV	M2A3	M
		Gun 25 MM M242 (G96797)	S
		Machine Gun (L92352)	S
		Radio (R45407)	C
		Launcher (L45740)	F
F81880	DCON APPR	M12A1	D
		1 Truck, 5T, M54A2C (X40794 or X40931) or M548 (D11049)	M
F90796	CFV	M3A3CFV	M
		Machine Gun 22MM M242 (G96797)	S
		Radio Set (R45407)	C
		Launcher (L45740)	F
G51840	GEN SET SMK	M157120GT, M15780GT, M157A212OG	D
		Truck, M1037 (T07543), M1097 (T07679)	M
G58151	GEN SMK	M56	D
		1 Truck, M1113 (T61630)	M
		1 Radio ANVRC90A (R67908) ANVRC-46 (G53001) (R67908, G53001)	C
G87229	GEN SMK	M58 D	
		1Radio (AN/VRC-87,88,89,90 or 91)	C
		1 M113A3 (C18234)	
H57505	HOW LT TWD	M119, M119A1	S
		Truck (T07679)	M

TABLE 6.4G	List of Ground Subsystems for DA Form 2406, continued

LIN	Noun Abbreviation	Subsystem	EOS Codes
H57642	HOW MD SP	M109A6	M
		Main Gun (L91975)	S
		Radio Set (R44795)	C
H76352	FLT CEN	TSC61LP, 61ALP, 61BLP	C
		Power Plant, MJQ10A, (P27819), MJQ40 (P42126)	P
		Truck, 2½T, M35A2 (X40009)	M
		1 Air Conditioner (A24455)	E
J04717	FSSP	Fuel System Supply Point	N
		2 Filter Separators, 350 GPM (H52087)	N
		2 Pump Assemblies, Flmbl Liquid (P97051)	N
		6 Tank Assemblies, Fabric Collapsible (V12552)	N
J30492	GEN SMK	2 M3A3 or 2 M3A4 (or 1 of each)	D
		1 Truck, M988/M1037 (T61494/T07543) or	M
		1 Truck, M151 (X60833), M1097 (T07679)	M
		Trailer (W95400)[5]	B
J81750	IFV	M2	M
		Main Gun 25 MM M242 (G96797)	S
		Radio Set (Q53001, Q56783)(R45407)	C
		Missile Launcher Assy TOW	F
K57392	HOW LT TWD	M101LT, M101ALT, M102	S
		Trk (T61494)	M
K57667	HOW MD SP	M109, M109A2, M109A3, M109A4, M109A5	M
		Main Gun	S
K57803	HOW MD TWD	M114, M114AI, M114A2	S
		Trk Cgo (X40968)	M
K57821	HOW MD TWD	M198	S
		Truck (X40968)	M
L36402	LDG CT CEN	TSQ71ALP, 71B	C
		Power Unit, PU678 (J50185)	P
		Truck, 2½T, M35A2 (X40009)	M
		Air Conditioner (A23684)	E
L36739	LCM	LCM8, LCM8MOD1, LCM8OD1SL, LCM8MOD1SLE	M
		Life Raft	N
		Radar Navigation	N
		HF Interface Unit	C
		Sonar, Digital Depth	N

5. ¼ Ton trailer required with M151 truck.

TABLE 6.4H	List of Ground Subsystems for DA Form 2406, continued

LIN	Noun Abbreviation	Subsystem	EOS Codes
L36876	LCU	LCU1466, LCU1466A, LCU1646, LCU1646MAR	M
		Life Raft	N
		Radar Navigation	N
		HF Interface Unit	C
		Sonar, Digital Depth	N
L43664	LNCH TNK C	M48A5, M60	M
		Radio Set (Q53001,R68010)	C
		60 Foot Brdg (C20414)	N
L67342	LCHR MCL	MK155, MK155M1, MK155M2, MK155M3	S
		Trailer (E02670, E02807)	B
L69306	RDO TML	TRC190V1, TRC190AV1	C
		Generator Set, PU751 (G37273) or	P
		PU797 (G42238)	P
		Truck, M1037 (T07543), M1097 (T07679)	M
		KYK13, KY57 (E98103, S01373)	K
L69374	RDO TML	TRC190V2, TRC190AV2	C
		Generator Set, PU751 (G37273), PU797 (G42238)	P
		Truck, M1037 (T07543), M1097 (T0 7679)	M
		KYK13, KY57, KG94A (E98103, S01373, T08971)	K
L69442	RDO TML	TRC190V3, TRC190AV3	C
		Generator Set, PU751 (G37273), PU797 (G42238)	P
		Truck, M1037 (T07543), M1097 (T07679)	M
		KYK13, KY57 (E98103, S01373)	K
L69510	RDO TML	TRC190V4, TRC190AV4	C
		Generator Set, PU751 (G37273), PU797 (G42238)	P
		Truck, M1037 (T07543), M1097 (T07679)	M
		KYK13, KY57 (E98103, S01373)	K
L70538	LAU ADV SYS	Generator Set Diesel Engine (G74575)	P
		Truck Tractor, 5 Ton (T61239)	M
		Semi trailer, Low Bed (S70027)	B
M04268	MGMT FAC	TSQ154A	C
		Truck, M1037 (T07543), M1097 (T07679)	M
		Generator Set, PU753 (G40744), PU798 (G42170)	P
M04941	MDS	TMQ31	C
		Power Plant, MJQ18 (P28015), MJQ37 (P42262)	P
		3 Trucks, 5T, M925 (X40931)	M

TABLE 6.41		List of Ground Subsystems for DA Form 2406, continued	
LIN	**Noun Abbreviation**	**Subsystem**	**EOS Codes**
M21948	MCS	TSQ138	C
		Generator Set, MEP114A (J36725), MEP815A (G74643) or 60KW on board M1015A1	P
		Carrier, M1015A1 (C10858)	M
		Air Conditioner 36KBTU (A24934)	E
M35941	METLOG ST	ANTMQ41	C
		Truck (T07679)	M
		Power Plant MJQ35 (P28083)	P
M57048	MIX PLT	KA60A, KA60	N
		Generator Set, MEP006A, (J38301) and MEP009B, (J40158)	P
P05439	OPER GRP	OL412TTC46B	C
		Generator Set, PU753 (G40744), PU798 (G42170) (Shared w/LIN S24750)	P
		Truck, M1037 (T07543), M1097 (T07679)	M
		KY57 (S01373), KY90 (S40395)	
P21220	PADS	USQ70	N
		Vehicle Truck, M1009 (T05028)	M
P60408	OPER GRP	413TTC47E	C
		Air Conditioner (A24463)	E
		Radio Set (R30963)	C
		Speech Security EQ (S01373)	K
		Truck (T07679)	M
P70292	OPER GRP	413TTC47B	C
		Generator Set, PU753 (G40744) (Shared w/LIN S24818)	P
		Truck, M1037 (T07543), M1097 (T0679)	M
P70360	OPER GRP	413TTC47C, 413TTC47DV1, 413TTC47DV2	C
		Generator Set, PU753 (G40744), PU798 (G42170)	P
		Truck, M1037 (T07543), M1097 (T07679)	M
		Trailer, M101A2 (W95537)	B
		KY57, KY90 (S01373, S40395)	K
R14148	RDR ST	TPQ36V1, 36V5, TPQ36V3	C
		Power Plant, MJQ25 (P42364), MJQ38 (P42330)	P
		2 Trucks, 2½T, M35A2 (X40009) or 2 Trucks, 5T, M813A1/M813A1WW (X40794/X40931)	M
R14216	RDR ST	TPQ36V7	C
		2 Generators MEP112 (G35981) or MEP813A (G74779)	P
		2 Trucks, 5T M1097 (T07679)	M

TABLE 6.4J **List of Ground Subsystems for DA Form 2406, continued**

LIN	Noun Abbreviation	Subsystem	EOS Codes
R14284	RDR ST	TPQ36V8	C
		Generator MEP112A (G35981) or MEP813A (G74779)	P
		2 Trucks, M1097 (T07679)	M
R33351	RDO ACC UT	TRC191AV1, TRC191AV2	C
		Generator Set, PU751 (G37273), PU797 (G42238)	P
		Truck, M1037 (T07543), M1097 (T07679)	M
		KYK13, KY57 (E98103, S01373)	K
R36854	RCV ST RDO	TRQ32, TRQ32V1	C
		2 Trucks, M1028A1 (T59414)	M
R38883	RCV ST RDO	TRQ37	C
		Truck Cargo, M1028 (T59414)	M
		Power Unit, PU620 (J47617)	
R39452	RDO TML ST	TRC173, 173A, 173B	C
		2 Generator Sets, MEP003 (J35825), MEP803 (G74711) or	P
		1 Power Unit, PU618 (J47480)	M
		Truck, 5T, M923 (X40794)	E
		2 Air Conditioners, 9 KBTU (A23955)	K
		KY57, 68, KG81 or KG94 (S01373, S64488, E03123, T64771)	
R39520	RPT ST RDO	TRC174, 174A, 174B	C
		2 Generator Sets, MEP003 (J35825) MEP 803A (G74711) or 1 Power Unit, PU618 (J47480)	P
		Truck, 5T, M923 (X40794)	M
		2 Air Conditioners 9KBTU (A23955)	C
		KY57, 68 (S01373, S64488)	K
R39588	RDO TML ST	TRC175, 175A, 175B	C
		2 Generator Sets, MEP003 (J35825), MEP803A (G74711) or	P
		1 Power Unit, PU618 (J47480)	M
		Truck, 5T, M923 (X40794)	E
		2 Air Conditioners, 9KBTU (A23955) KY57, 68K (S01373, S64488)	
R41282	RECON SYS	M93A1	D
		Machine Gun (L92352)	S
		Radio Set (R44863) or (R67908)	C
R50544	REC VEH LT	M578	M
		Machine Gun .50 CAL (L91975)	S
		Radio Set (Q56783)	C
R50681	REC VEH MD	M88A1	M
		Machine Gun .50 CAL S (L91975)	S
		Radio Set (Q53001) R44795, R44863, R45339, R45407, R67228, R67262, R44931, R67976, R68078, R45475	C

TABLE 6.4K	List of Ground Subsystems for DA Form 2406, continued		

LIN	Noun Abbreviation	Subsystem	EOS Codes
R50885	REC VEH FT	M88A2	M
		Machine Gun .50 Cal (L91975)	S
		Radio Set (R45271) (R67908) R67228, R67262, R44931, R67976, R45475	C
R78116	RPT ST RDO	TRC138A, TRC138B , TRC138C, 2 Generator Sets, MEP003A	C
		(J35825), MEP803A (G74711) or 1 Generator Set, PU631 (J46396)	P
		Truck, 5T, M923 (X40794)	M
		2 Air Conditioners, 9KBTU (A23955)	E
		KG57, 68 (S01373, S64488)	K
R92967	RDO TML ST	TRC170V2	C
		2 Generator Sets, MEP005A (J36109) or MEP805A (G74575)	P
		Truck, 5T M923 (X40794)	M
		Truck, 2½T, M35A2 (X40009)	M
		KY68 (S64488), KG94 (T64771)	K
R92996	RDO TML ST	TRC145BV1,TRC145BV1	C
		Air Conditioner, 18KBTU (A26271)[6]	E
		Power Unit/Generator Set, PU625 (J46252)	P
		Truck, 1¼T, M885 (X39441) or M1028 (T59414)	M
		KG27 (L22987)	K
R93035	RDO TML ST	TRC170V3	C
		Power Plant/Generator Set, G42170, (J35825)	P
		2 Trucks, M1097 (T07679)	M
		KY 68, (S64488), KG 94, (T64771)	K
S24750	SWTCH GRP	305TTC46, 305TTC46A, 305TTC46B	C
		Generator Set, PU753 (G40744), PU798 (G42170) (Shared w/LIN P05439)	P
		Truck, M1037 (T07543), M1097 (T07679)	M
		KG94A (T08971)	K
S24818	SWTCH GRP	0N306TTC47, 47A, 47B	C
		Generator Set, PU753 (G40744), PU798 (G42170) (Shared w/LIN P70292)	P
		Truck, M1037 (T07543), M1097 (T07679)	M
		KG94A (T08971)	K
S25379	SENS	TTC48V2, 48AV2, 48BV2	C
		Generator Set, PU753 (G40744), PU798 (G42170)	P
		Truck, M1037 (T07543), M1097 (T07679)	M
		KYK13, (E98103) KG94A, (T08971) KY57, (S01373), KY 90 (S40395)	K

6. Count the air conditioner subsystem only when it is authorized and mission essential in your area.

		TABLE 6.4L List of Ground Subsystems for DA Form 2406, continued	
LIN	Noun Abbreviation	Subsystem	EOS Codes
S25447	SM EXT	Air Conditioner, 18000BTU (A24463)	E
		Generator Set, Diesel Engine (PU753-M) (G40744)	P
		Truck Utility HMMWV (T07679)	M
		Generator (TSEC/KG94A (T08971)	P
S25515	SENS	TTC48V1, TTC48AV1, 48BV1 C	
		Generator Set, PU753 (G40744), PU798 (G42170)	P
		Truck, M1037 (T07543), M1097 (T07679)	M
		KYK13, (E09103), KG94A, (T08971), KY57, (S01373), KY 90 (S40395)	K
S34963	SAT COM TM	TSC93BV1, TSC93B	C
		2 Generator Sets, PU753 (G40744), PU7798 (G42170)	P
		2 Generator Sets MEP003A (J35825), MEP803A (G74711)	P
		2 Trucks, 2½T, M35A2C (X40077)	M
		2 Trucks, 1¼T, M1028 (T59414)	M
		2 Trucks, 5T, M923 (X40794)	M
		2 Trucks, 5T M1097 (T07679)	M
S37228	SWTCH GRP	306TTC47C	C
		Truck (TO7543), (T07679)	M
		Generator Set PU753, (G40744), PU798 (G42170)	P
		Trailer (W95537)	B
		HGF96, (Z92634), KGX93A, KG112, (Z25051), KG194A (Z92634) (Z25051) (T08971)	K
S38172	SENS	TTC48CV	C
		Truck (T07543) (T07679)	M
		Generator Set (G40774), (G42170)	P
		Trailer (W95537)	B
		KG194A, KY90, KY57 KYX15, (T08971) (S40395) (S01373) (N02758)	K
S44664	CNTRL GRP	OL414TYQ35	C
		Generator Set, PU751 (G37273)PU797 (G42238)	P
		Truck, M1037 (T07543), M1097 (T07679)	M
S44732	CNTRL GRP	OL416TYQ35	C
		Generator Set, PU753 (G40744), PU798 (G42170) (Shared w/LIN S44914)	P
		Truck, M1037 (T07543)	M
S44914	CNTRL GRP	OL415TYQ35	C
		Generator Set, PU753 (G40744), PU798 (G42170) (Shared w/LIN S44732)	P
		Truck, M1037 (T07543), M1097 (T07679)	M

	Noun		
LIN	**Abbreviation**	**Subsystem**	**EOS Codes**
S78466	SAT COM TM	TSC85BV1 TSC85A	C
		2 Trucks, 2½T, M35A2C (X40077)	M
		2 Trucks, 5T, M923, (X40794)	M
		2 Generator Sets, PU405A (J35492), PU802 (G53788)	P
S78717	SW GP	Truck Utility, Heavy HMMWV (T07679)	M
		Trailer Cargo 1¼T (T95924)	M
		Generator Set 10KW PU-79 (G42170)	P
T10138	SP EQ MNT	993, AVNC6217, CMU3, CMU5, MILS45855, SECM1975	T
		Truck	M
T10275	SP EQ ELEC	FSVAN15777, FSVAN1959, MILS52330, SER1961, SER1968, SER1976, SER197881, SER1982,CBL05	T
		Semitrailer	B
T10412	SP EQ ELEC	ELECREP, MILS52377, SEER1963, SEER1968	T
		Truck, 5T	M
T10549	SP EQ ELEC	MED1952, ENG 4359, MILS45538, SGPRSMD, SGPRSM61, SGPRSM68	T
		Generator	P
		Truck, 5T	M
T13152	SP EQ ORG R	ENG40, MEDL1954, MEDL1956, MILS45537 SEORL118, SEORL66, SMGPR61, SOUTHWEST, SEORTM	T
		Truck	M
T13168	TNK CBT FT	M1A1	M
		Main Gun	S
		1 Machine Gun, coax 7.62MM (L92352)	S
		1 Machine Gun, .50 CAL (L91701)	S
		Radio Set (R45407), (R44863, R67160)	C
T13169	TNK CBT FT	M60A3TTS	M
		Main Gun	S
		1 Machine Gun, coax 7.62MM (L92352)	S
		Machine Gun, .50 CAL (L91701)	S
		Radio Set (Q53001, Q56783)	C
T13305	TNK CBT FT	M1A2	M
		Main Gun	S
		1 Machine Gun, coax 7.62MM (L92352)	S
		1 Machine Gun, 50 Cal (L91975)	S
		Radio Set (R45407)	C
T13374	TNK CBT FT	M1, M1IP	M
		1 Machine Gun, 7.62MM (L92352)	S
		1 Machine Gun, .50CAL (L91701)	S
		Radio Set (R44659, R44795)	C

TABLE 6.4M **List of Ground Subsystems for DA Form 2406, continued**

TABLE 6.4N		List of Ground Subsystems for DA Form 2406, continued	

LIN	Noun Abbreviation	Subsystem	EOS Codes
V12141	TNK PMP UT	MDL1800,MD 2938, MD1151 ENG2519, HLND2000, ORRBL100, BOW36W50, ALTECH, 13217E7100, 13217E7130, 126ETP	N
		Truck, 5T	M
V13101	TNK CBT FT	M60A3,	M
		Main Gun	S
		1 Machine Gun, coax 7.62MM (L92352)	S
		1 Machine Gun, .50 CAL (L92112)	S
		Radio Set (Q53001, Q56783)	C
V57504	TML TG	TSC58, TSC58A,	C
		Air Conditioners, 9KBTU (A23828)	E
		Generator Set, PU619 (J42100)	P
		Truck, 2½T, M35A2 (X40009)	M
		KW7 (H02300)	K
W35417	WTR PURIF	ROWPU600	N
		WSPES1, WPES10 Tank Assy 3000 Gal (T19033)	N
		Generator Set (J35835)	P
		Trailer (W95811)	B
		Pump (P92030, P91756 or P44549)	N
W47225	WTR PURIF	ROWPU3000, ROWPU1	N
		Tank Assy 3000 GAL (T19033)	N
		RAW Water Pump (P92030)	N
		Generator Set (J38301)	P
		Truck (X59463)	M
		Trailer (S70027)	B
Y35486	WPE 1500	1500GPH	N
		Tank (V14881)	N
		Pump Centrifugal (P92030)	N
		Generator Set (J49398)	P
		Trailer (W95811)	B
		Truck (X40009)	M
Y36034	WPE 3000	3000GPH	N
		Tank (V15018)	N
		Pump Centrifugal (P92030)	N
		Generator Set (J38712)	P
		Trailer (S70027)	B
		Truck (X59463)	M

TABLE 6.5			List of Reportable Aircraft Systems for DA Form 1352			
ECC	LIN	EIC	Nomenclature Series	Noun Abbreviation	Model Design	NSN
AF	29744	SCB	Airplane	APLN	12C	1510010703 661
AF	A29812	SCC	Airplane	APLN	C12D	1510010879 129
AF	A29880	SAA	Airplane	APLN	C23B	1510994955 760
AF	A29880	WG5	Airplane	APLN	C23B Plus	1510014181 848
AF	A30062	SCF	Airplane	APLN	C12F	1510012355 840
AF	A30312	SCE	Airplane	APLN	C12L	1510012652 043
AF	A30989	SVB	Airplane	APLN	UV18A	1510010111 462
AF	Z04378	SCG	Airplane	APLN	RC12G	1510012152 942
AF	Z04549	SCD	Airplane	APLN	RC12D	1510011318 262
AR	A21633	ROC	Helicopter	HCPTR	OH58D	1520011255 476
AR	H28647	RHA	Helicopter	HCPTR	AH64A	1520011069 519
AR	H29762	RAD	Helicopter	HCPTR	AH1P	1520011684 259
AR	H30517	RCD	Helicopter	HCPTR	CH47D	1520010883 669
AR	H30616	RSB	Helicopter	HCPTR	EH60A	1520010820 686
AR	H30766	RSC	Helicopter	HCPTR	MH60K	1520012824 051
AR	H31110	ROB	Helicopter	HCPTR	OH58C	1520010204 216
AR	H31872	RUE	Helicopter	HCPTR	UH1V	1520010434 949
AR	H32361	RSM	Helicopter	HCPTR	UH60L	1520012984 532
AR	H32611	RTB	Helicopter	HCPTR	TH67A	1520013853 844
AR	H44644	RAF	Helicopter	HCPTR	AH1F	1520011684 260
AR	H44712	RAE	Helicopter	HCPTR	AH1E	1520011922 478
AR	H46150	RCE	Helicopter	HCPTR	MH47E	1520012823 747
AR	H48918	RHB	Helicopter	HCPTR	AH64D	1520013558 250
AR	K29694	RAA	Helicopter	HCPTR	AH1S	1520005049 112
AR	K31042	ROA	Helicopter	HCPTR	OH58A	1520001697 137
AR	K31795	RUA	Helicopter	HCPTR	UH1H	1520000877 637
AR	K32293	RSA	Helicopter	HCPTR	UH60A	1520010350 266

TABLE 6.6	List of Reportable Missile Systems for DA Form 3266-1	

ECC	LIN	Nomenclature
BL	C40746	JOINT TACTICAL GROUND STATION (JTAGS)
BP	011111	PATRIOT FIRING BATTERY
BN	F57713	AVENGER
BL	G92997	SENTINEL, RADAR SET ANMPQ64
BM	L60078	LIGHT SPECIAL DIVISION INTERIM SENSOR (LSDIS)
		Land Combat Systems
CF	C12155	GROUND VEHICULAR LASER LOCATOR DESIGNATOR (GVLLD) M981, A3
CC	E56896	TOW 2, IMPROVED TOW VEHICLE (M901A1, A3)
CG	L44894	MULTIPLE LAUNCH ROCKET SYSTEM
CC	L45740	TOW 2, HMMWV (14440-01-411-8942, 1440-01-410-8165)
CC	T24690	Target Acquisition
CF	T26457	GROUND VEHICULAR LASER LOCATOR DESIGNATOR (GVLLD)
CZ	T92961	BASE SHOP TEST FACILITY ANTSM191V3
		Land Combat Equipment
CD	N23721	NGT VIS SGT DRAGON
CD	C60750	CMD LNCH UNIT JAVELIN
CD	W80715	TRACKER DRAGON

REMEMBER . . .

Reward for a
job well done!

CONCLUSION

The Army is transforming and modernizing in order to accomplish its 21st-century missions. The Army's combat and combat support forces must generate combat power to accomplish those missions. The Army can best generate combat power when its equipment meets the Army Maintenance Standard. Army maintainers, in support of operators, crew, and other users, have a key role in sustaining the means of combat power. As an Army leader, you must make sure that this happens.

Key Words

supply
maintenance
logistics
property book
preventive maintenance checks and services (PMCS)
command maintenance program
test, measurement, and diagnostic equipment (TMDE)
maintenance assistance and instruction team (MAIT)
unique item tracking (UIT)
not mission capable (NMC)

Learning Assessment

1. Explain the three sections of DA Form 5988-E.
2. Explain the purpose of the not-mission-capable (NMC) report.
3. Identify the PMCS responsibility of key unit personnel.
4. Identify the purpose of the command maintenance program.
5. Define PMCS workflow.

References

AR 700-138, *Army Logistics Readiness and Sustainability*. 26 February 2004.

DA PAM 750-1, *Commanders Maintenance Handbook*. 2 February 2007.

DA PAM 750-8 *The Army Maintenance Management Systems (TAMMS) User Manual*. 22 August 2005.

Field Manual 4-30.3, *Maintenance Operations and Procedures*. 28 July 2004.

Section 7

FINANCIAL MANAGEMENT

Key Points

1 **Planning Individual and Family Finances**

2 **Your Leave and Earnings Statement**

3 **Resolving Pay Inquiries**

4 **Support Agencies and Programs**

There is no dignity quite so impressive, and no independence quite so important, as living within your means.

President Calvin Coolidge

Introduction

In today's financial environment, having a big salary and living beyond your means do not determine personal success. Personal success more often results from careful financial planning, careful spending, and careful saving. Over the years, government, big businesses, and many individuals have adopted a buy-now, pay-later attitude as a way of life. Living beyond one's means has become a norm, causing many people to end up broke or filing for bankruptcy—all because they've poorly managed their personal finances. This way of thinking is not in the best interest of any Soldier or Soldier's family members.

As you prepare to graduate and become an Army officer, "a leader of Soldiers," you will need to take a good look at your personal finances. You should know how to manage your income, spending, and debt; where to obtain reliable financial information; and how to enroll in the Army's various savings and investment programs. You should also be familiar with the several support agencies (military and civilian) that can help Army officers and their families attain their financial goals.

This section will review some basics of financial survival and the Army's resources to help you with financial management and investment. Not only must you be familiar with these resources for your own day-to-day financial management—you will need to know them well enough to help your Soldiers (and their families) with their own financial questions and difficulties. By doing so, you will be able to strengthen each Soldier's ability to fulfill his or her financial responsibilities and behave as an informed consumer.

Remember that financial issues can and will affect your Soldiers' operational readiness and job performance. One of your responsibilities as an effective leader is to have a financial plan yourself and help your Soldiers to develop their own.

Remember that Soldiers from different backgrounds may have gaps in their knowledge and understanding of finances and how to mange their money. Some may never have had as much money as they are now earning in the Army. Some may never have had a bank account. Therefore, it's important to reinforce some basic financial principles. Listen to your Soldiers and take their questions seriously. Answer them fully, and treat all Soldiers with respect and dignity, while protecting their privacy.

Planning Individual and Family Finances

A Soldier who can demonstrate reasonable financial management is a Soldier better prepared for the unexpected. The better Soldiers are prepared, the better able they are to take care of their families—and thus the better they can focus on their jobs. Financial planning for individual Soldiers and their families covers a wide range of topics. Below are some of the areas a Soldier can start with to improve his or her financial readiness.

Never write checks on your account until you know that you have enough money in the account to cover the checks.

Financial Planning

Personal financial management is for everyone, not just people who have a lot of money. Think of it in terms that make it real to you and your goals will begin to take shape. Making the most of your money and buying wisely are things you can do *now*. Living well doesn't happen all at once. As with any goal, you need self-control and patience. As a leader, this is where you can help your Soldiers. Think about how you spend your money. Some tips for spending wisely include:

- Create a budget and stick to it
- Pay yourself first—setting aside a small percentage (say 15 percent) of your pay in a savings account
- Save regularly—an individual retirement account (IRA) and the government's Thrift Savings Plan (TSP) are good places to start
- Establish an emergency fund equal to three to six months' living expenses
- Buy wisely; be a smart consumer—look for specials and clip coupons, and avoid impulse purchases
- Save enough money to live comfortably now
- Save enough so you won't have to worry about money in old age.

Recognizing Financial Trouble

Most people don't get into financial trouble intentionally—they do so thoughtlessly. They say things like, "I didn't know the payments were due the first *day* of the month!" or "I deserve this!" Consider some of the things that can happen if you *don't* handle your money responsibly:

- *Rising debt*—Debt can creep up almost unnoticed and become unmanageable.
- *Lost possessions*—If you don't pay bills on time, you will lose things, like your car, your furniture, or your home. Any loan "secured" by a physical object, like a car, gives the loan company the option of *repossession* if that loan is not repaid on time or in full. Often, you'll still have to make payments on the items taken back.
- *Loss of credit*—If your creditors report your nonpayment to a credit bureau, you may not be able to borrow more, charge more, or sign new contracts or leases. Banks, credit card companies, stores, and landlords all have access to your credit report.
- *Pay garnishment*—Retailers in towns with military bases know you get a regular check and that they can easily **garnishee**. They also know your CO hates dealing with financial problems. Your commanding officer will probably be the *first* to know about your financial troubles.
- *Lost security clearance*—Your security clearance may immediately be jeopardized.
- *Lost privileges*—You may lose chances for promotions, duty assignments, and other privileges.
- *Inability to Re-enlist*—You may not be able to re-enlist an enlisted Soldier with financial issues.
- *Discharge from the service*—For serious, continuing problems, you can be discharged from the service with no recourse. Then you'd have no job—but you'd still have bills.

Take ownership of your future, don't let rising debt spin out of control.

garnishee (garnish)

to seize by garnishment, a legal proceeding in which a debtor's money or property is applied to pay a debt

Debt Management

If you find yourself in the position of considering loan consolidation, you may be headed for financial difficulty. The following are danger signs:

- You use more than 20 percent of your take-home pay for credit payments (excluding your home mortgage)
- You have one or more loans from a lending company or companies at 20 percent or more interest
- You screen your phone calls because bill collectors are calling frequently
- You use credit cards impulsively
- You routinely use the overdraft protection in your checking account
- You don't pay your bills on time
- You pay only the minimum amount due on your charge accounts each month
- Your regular budget plan includes using advance check-cashing and payday loans
- Your car loan is financed at 12 percent interest or higher.

Banking and Credit

Checking Accounts

A checking account is always a good idea, because it gives you virtually instant access to your money. The terms of checking accounts can be very different from bank to bank, so it pays to shop around. Some checking accounts have a minimum balance requirement, monthly maintenance fees, and overdraft fees. Some banks, however, offer accounts with no fees if you sign up for direct deposit (such as SUREPAY), and some actually pay you a small amount of interest each month on your average daily balance. Check with your bank or lending officer for more information. Since you will most likely move several times during your Army career, you will also want to consider banks that are located near a variety of Army installations, or banks that make it easy for you to conduct business via mail, telephone, or the Internet.

You should never write checks for more than the money you have in the account. Be sure to balance your account once a month, either using the bank statement you receive in the mail or your bank's website. If you use an ATM or debit card, remember that it is tied to your checking account, so record your electronic transactions in your checkbook register.

Debit Cards

Debit cards look just like credit cards, right down to the MasterCard or Visa logo, but they are very different. When you use a debit card, you make a withdrawal from your checking or savings account. Just as when you write a check, you shouldn't use a debit card unless you have money in your account to cover the withdrawal. Unlike when you write a check, the withdrawal from your account is immediate (because it's electronic), so you must always have an accurate idea of your account balance when you're using your debit card.

Credit Cards and Debt

Credit cards are a wonderful convenience, but they have also led incautious people into crushing levels of personal debt. It's important to understand that a credit card is basically a loan—only the interest rate on a credit card is generally much higher than for other types of loans. The key to wise credit card use is to *pay off the balance each month* and avoid costly finance charges. When you pay only the minimum monthly charge, your debt doesn't go down, *it often goes up*, because the credit card company charges you interest on the unpaid balance. Thus the total you owe could rise more than the minimum charge you paid, leaving you owing more than you did last month, even though you made a payment.

So if you're not careful, your unpaid balance can quickly get out of control. Consider the example of purchasing a $1,500 widescreen HDTV. You can save or set aside $250 a month and delay your purchase for six months when you can pay cash. Or, you can succumb to the temptation of instant gratification and use your credit card. What is the true cost of using credit if your credit card's annual rate is 18 percent? If you made the $35 minimum monthly payment, your $1,500 TV would end up costing you $2,420 and would take you nearly six years to pay off. Even if you paid $100 each month your TV would cost you $1,712, or $212 more, and it would take you 18 months to pay off.

Credit cards have various interest rates and spending limits. Going over your spending limit or not paying your minimum charges can cause problems for you in the Army. Sometimes credit companies will actually contact the unit commanders of Soldiers who owe them money. Failure to pay a just debt is a violation of the Uniform Code of Military Justice and can result in administrative separation.

Command Financial NCO (CFNCO)

The Command Financial NCO (CFNCO) program was created to enhance and maintain mission readiness and quality of life by helping Soldiers and their family members achieve personal financial readiness and deployability using sound money-management and consumer skills.

Each battalion-size element has a mature and financially stable CFNCO (E-6 thru E-8) who trains, organizes, implements, and supervises the program while serving as the battalion commander's principal adviser on policies and matters related to personal financial readiness and local consumer affairs.

Each CFNCO provides the following services:

- Conducts financial evaluations
- Provides budget counseling and advice to Soldiers and their family members
- Screens and counsels all Army Emergency Relief (AER) referrals
- Refers Soldiers and their family members to appropriate resources or agencies
- Presents financial-readiness and consumer training as part of the command program
- Disseminates financial and consumer information within the command
- Schedules unit financial-management training with the Financial Readiness Program
- Provides emergency food-locker referrals to Soldiers and family members
- Attends battalion predeployment briefings to describe resources available to family members
- Attends Army Family Team Building (AFTB) sessions as an available resource
- Provides financial and consumer benefits to all newly assigned personnel
- Assists the unit commander in emphasizing consumer prevention and education rather than crisis management.

Insurance

Servicemembers' Group Life Insurance (SGLI)

optional government-sponsored life insurance available to Soldiers

Military servicemembers can purchase **Servicemembers' Group Life Insurance (SGLI)**. This is an extremely low-cost, term life insurance policy that will pay up to $400,000.00 upon a servicemember's death. Although SGLI payments are deducted from your pay, you do not pay the whole premium. The government pays, or subsidizes, part of the premiums, making the cost of this insurance very low for you. You also have the option of selecting an amount of SGLI coverage smaller than standard, thereby lowering your premiums. You can start or change this insurance coverage at your local finance office.

Budgeting and Investments

Soldiers must learn to stick to a budget that allows them to live within their means—to spend only what they can afford, or better yet, less. Go beyond those means and you can find yourself experiencing the danger signs of financial difficulty or possible bankruptcy. Living within your means is the best way to avoid financial trouble. You get a start by paying off your debts and investing in your future.

Individual Retirement Accounts (IRAs)

An individual retirement account (IRAs) is an excellent way to put aside money each week or each month to prepare for retirement. IRAs are available through most financial institutions. You can have as many separate IRAs as you want, but the law limits the amount you can deposit each year to $5,000 for those age 48 and below. You can even transfer money directly from one IRA to another.

There are two different types of IRAs—*traditional* and *Roth*. They differ in several ways:

- You pay into a traditional IRA using *pre-tax* dollars. That is, you pay no income tax on the money you invest in a traditional IRA (unless you have a very high income)—instead you pay the tax as you withdraw the money during retirement. In a Roth IRA, you pay with *after-tax dollars* and pay no income tax on the gain when you withdraw money.
- Roth IRAs offer no withdrawal penalties if the account has been open for five years and the taxpayer is at least $59\,{}^1/_2$ years old or is using the withdrawal for college or to buy a first home.
- Unlike traditional IRAs, Roth IRAs have no requirement that you begin withdrawing by age $70\,{}^1/_2$.

You may wish to consult a representative of your financial institution to help you decide which type of plan is best for you.

Thrift Savings Plan

Starting a **Thrift Savings Plan (TSP)** is another way of planning for your future. Here are five key points every Soldier should know about a TSP:

- It's a retirement savings and investment plan that has been available to government civilians since 1987
- The purpose is to provide retirement income, much like a civilian 401(k) plan
- A TSP allows Soldiers to set aside a portion of their pay in a special retirement account administered by the Federal Retirement Thrift Investment Board
- Money invested comes from pre-tax dollars—lowering your taxable income—and earnings are not taxed until you withdraw them in retirement
- Participation in the TSP is optional, not automatic.

> **Thrift Savings Plan (TSP)**
>
> *a retirement savings and investment plan available to government employees, including Soldiers*

Military retirement is a *benefit program*—your pension is based on your rank and time in service. On the other hand, TSP is a *contribution* plan—the balance in your TSP account depends on how much you contribute and the earnings on your savings. Soldiers may sign up for the TSP at any time.

Soldiers may also contribute all or any whole percentage of any special or incentive pay they receive. Contributions are deducted per pay period; you can't send checks directly to the plan. To contribute from a bonus, you must already be contributing from your basic pay. Any contributions from a bonus must be deducted by the finance department at the time the bonus is paid. Soldiers may contribute to an IRA in addition to contributing to the TSP.

The TSP is a *long-term retirement* plan. It gives Soldiers an opportunity to save and build a nest egg over the long haul, along with their retirement pension. Since the government controls it, it will always be a conservative and stable investment that will see generally steady slow growth, depending on the stock and bond markets. The TSP is not a short-term savings account, but for the Soldier who has time to allow the account to grow, this is a great deal.

Planned Deployment

In today's military operating environment, almost every Soldier can count on finding himself or herself facing overseas deployment. Each Soldier deals with deployment in his or her own way. But you can help your platoon members by being proactive and providing them a "Predeployment Checklist." Using the checklist to ensure they have their affairs in order can help Soldiers plan and ease some of the stress on them and their families. Table 7.1 shows an example of a predeployment checklist.

TABLE 7.1	Predeployment Checklist

PERSONAL/FAMILY

- ID Cards
- DEERS Enrollment
- Wills
- Power of Attorney
- Family Care Plan (Single/Dual Career)
- Insurance Policies
- Birth Certificates
- Marriage Certificates
- Adoption Papers
- Immunization Records
- Citizenship Certificates
- Passports
- Personal Property Inventory
- Fingerprints/ID-A Kit for Children
- Divorce Decrees
- Naturalization Papers
- Discharge Papers (DD 214)
- Vehicle/Boat Registration

FINANCIAL/BUSINESS

- Bonds
- Certificates of Deposit
- Stocks
- Household Inventory
- Securities Account Records
- Deeds
- Mortgage Papers
- Vehicle Titles
- Real Estate Appraisals
- Real Estate Sales Records
- Checking/Savings Account Information

Your Leave and Earnings Statement

Your Leave and Earnings Statement (LES) is a comprehensive statement of your leave and earnings showing your entitlements, deductions, allotments (fields not used for Reserve and National Guard members), leave information, tax-withholding information, and Thrift Savings Plan (TSP) information. You can find your most recent LES 24 hours a day on myPay: *https://mypay.dfas.mil/mypay.aspx*. You should review your LES each pay period for accuracy.

Military Pay Entitlements and Allowances

As a newly commissioned Army officer, you will need to know the various acronyms and names used to identify the many entitlements and allowances on a Soldier's LES. Table 7.2 gives a list of the most common allowances found on the LES. Two of the most important are the **Basic Allowance for Housing (BAH)** and the **Basic Allowance for Subsistence (BAS)**. You can find more information on each entitlement or allowance on the **Defense Finance and Accounting Service (DFAS)** website at *http://www.dfas.mil*.

Basic Allowance for Housing (BAH)

money the Army pays Soldiers for housing, usually when adequate government quarters are not provided

Basic Allowance for Subsistence (BAS)

a food allowance paid to all officers and enlisted Soldiers

Defense Finance and Accounting Service (DFAS)

the Defense Department's accounting and finance agency

TABLE 7.2A	Allowances and Entitlements

TYPE OF ENTITLEMENT	DESCRIPTION
1. Basic Allowance for Housing (BAH)	The intent of BAH is to provide uniformed service members on permanent duty within the 50 United States accurate and equitable housing compensation based on housing costs in local civilian housing markets. BAH is payable when government quarters are not provided. The types of BAH are: ***BAH With*** or ***Without Dependents***, ***BAH-partial*** (a member without dependents who is assigned to single-type quarters or is on field or sea duty, and is not authorized to receive a BAH or OHA at the full rate, is authorized BAH-partial), ***BAH-Diff*** (which applies to service members who pay child support), ***BAH-Transit*** (a temporary housing allowance paid while a member is in a travel or leave status between duty stations, provided the member is not assigned to government quarters), and ***BAH-RC*** (Reserve Component Rate, which is the housing allowance authorized for a Reserve Component member called or ordered to active duty for 30 or fewer days, except for a Reserve Component member called to active duty for a contingency).
2. Basic Allowance for Subsistence (BAS)	For enlisted Soldiers, this is commonly known as *separate rations*. All enlisted Soldiers receive full BAS and pay for their meals, except while they are on deployment or on field or sea duty, when they may lose the food allowance because they receive rations. Officers receive BAS and pay for all their meals. Officers on field or sea duty continue to receive BAS and continue to pay for all their meals.
3. Clothing Allowance	*Initial Clothing Allowance* Both officers and enlisted members of the Armed Forces are entitled to an initial clothing allowance. Officers, however, are only entitled to the allowance once except in the situations noted below: Upon first reporting for active duty (other than for training) for a period of more than 90 days. Upon completing at least 14 days of active duty or active duty for training as a member of a Reserve Component. Upon completing 14 periods of inactive-duty training as a member of the Ready Reserve. (Each period must be of at least two hours duration). Upon reporting for the first period of active duty required of a member of the Armed Forces Health Professions Scholarship Program. NOTE: Upon transfer to another Reserve Component that requires a different uniform, a Reserve officer may receive another initial uniform allowance. Regular officers may not receive this allowance when transferring to another military service. *Cash Clothing Replacement Allowance* This allowance is payable only to enlisted members annually following the initial clothing allowance on the anniversary month. This is to replace uniforms based on normal wear and tear.

TABLE 7.2B	Allowances and Entitlements, continued

TYPE OF ENTITLEMENT	DESCRIPTION
	Extra Clothing Allowances
	Extra Clothing Allowances are additional to the other two and do not affect them. These allowances are for situations in which a member may need an additional uniform or is required to have civilian clothing to perform his or her duties.
	Military Clothing Maintenance Allowance
	This allowance is for replacement and maintenance of military items during and after three years of active duty.
	If a member has a break in military service, he or she will start over with the initial clothing allowance upon returning to the service. He or she will then receive a Cash Clothing Replacement Allowance and Military Clothing Maintenance Allowance as applicable.
4. Family Separation Allowance (FSA)	To receive the FSA you must meet the following criteria:
	You must have dependents
	You must be on temporary duty away from the permanent duty station for a continuous period of more than 30 days and the dependents do not live at or near the temporary duty station.
5. Cost-of-living Allowance (COLA)	COLA is a cash allowance intended to enable an equitable standard of living in areas where cost of living is unusually high. Members permanently assigned to designated areas receive this entitlement. If the cost of living in the area where the member is assigned is the same or lower than average in the United States, COLA is not authorized.
	Soldiers deploying from an area not qualifying for COLA are not entitled to COLA while deployed, even if the deployed location is a designated COLA area.
6. Deployed Entitlements/ Special Pays	If a service member is deployed, he or she may be eligible to receive any of the following: • Per diem • Family Separation Allowance • Basic Allowance for Subsistence (meals) • Basic Allowance for Housing (based on home station) • Cost of Living Allowance (if eligible prior to deployment) • Hardship Duty Pay-Location • Hostile Fire/Imminent Danger Pay • Combat Zone Tax Exclusion • Savings Deposit Program • Special Leave Accrual • Tax Filing Extension • Special Extension Entitlements (for extended periods of deployment).
7. Overseas Housing Allowance (OHA)	OHA is a monthly allowance paid to service members assigned to a permanent duty station outside the continental United States (OCONUS), except Hawaii and Alaska, who are authorized to live in private housing.

TABLE 7.2C	Allowances and Entitlements, continued

TYPE OF ENTITLEMENT	DESCRIPTION
8. Dislocation Allowance (DLA)	The purpose of DLA is to partially reimburse a member, with or without dependents, for the expenses incurred in relocating the member's household on a permanent change of station (PCS), housing moves ordered for the government's convenience, or incident to an evacuation. This allowance is in addition to all other allowances authorized and may be paid in advance.
9. Move-in Housing Allowances (MIHA)	Move-In Housing Allowances are paid in lump-sum supplemental payments when a member becomes eligible for OHA and incurs occupancy-related expenses. There are three types of move-in allowances: miscellaneous, rent, and security (MIHA).
10. Geographical Bachelor	A Geographic Bachelor is a servicemember (Airman, Marine, Sailor, or Soldier), collecting BAH (at the with dependents rate), authorized to be accompanied by dependents, eligible for family housing, who for personal reasons other than availability of housing at the permanent duty location, is not accompanied by dependents.

Resolving Pay Inquiries

From time to time in every unit, there will be Soldiers facing financial and pay issues. As a leader of Soldiers, you may find yourself involved in helping resolve such pay issues. Some pay inquiries may require research to find the correct answer to the Soldier's question. It's important to address a Soldier's pay issue in a timely manner, preferably within 48 hours.

The procedures for Soldiers in the active Army, the Reserve, or the National Guard differ slightly, depending on location and the organization's standing operating procedures (SOP). The steps listed below, however, will get you started in helping the Soldier.

First, begin with the chain of command, starting with the Soldier's unit. Work the issue as much as you can—many times, it can be resolved or answered within the company. Take any pay inquiries that you can't resolve at the unit level to the next level in the chain of command—but only after you've exhausted all possibilities at the lowest level.

Second, have the Soldier fill out the required document to record his or her concerns. For active Army, use AR 37-104-4, DA Form 2142 (Pay Inquiry); for Reserve Component Soldiers use USARC Reg 37-1, USAR Form 27-R (Pay Inquiry). You can find these forms at *http://www.usuhs.mil/bde/forms/PayInquiry2142.pdf* and *http://www.usarc.army.mil/ 377TSCBAK/SoldierSupport/Finance/Forms/PDF/usarc27r.pdf*.

Third, make sure the Soldier states his or her concerns in the designated section of the form. Both the DA Form 2142 and the USAR Form 27-R are one-page documents with multiple sections. The Soldier can sign the form; but be sure to comply with your chain of command's SOP. The Soldier should complete each section, and the authorized individual should sign the form. Along with the form, furnish all the documents needed to support the Soldier's complaint. Check to see what your unit's SOP requires.

Fourth, once the Soldier has completed the appropriate sections of the form and obtained the required signatures, forward it to the Soldier's administrative section (next level up) for processing. At this point, the form will be logged for record purposes and the issue researched to find an answer for the Soldier and the chain of command. Depending on how complex the issue is, resolving it may require higher-level review and in-depth research. If getting an answer will take some time, the local administrative section (S-1, RSC/DRC) should attempt to keep the Soldier informed through the chain of command.

Finally, if all efforts have been made to resolve the issue through the chain of command and the Soldier cannot get a response to his or her concerns, the last step is to file a complaint with the installation inspector general.

Support Agencies and Programs

Many support programs are available to Soldiers and their families who may be in difficulty. Once you become familiar with the services that various organizations offer, you can help your Soldiers find the most appropriate source of support for their particular needs.

Army Emergency Relief

Army Emergency Relief (AER) is a nonprofit organization that provides emergency financial assistance to active and retired Soldiers and their dependent family members in times of distress. AER is authorized to provide emergency financial assistance in cases of:

- nonreceipt of pay
- loss of funds (the loss must be reported to the military or civilian police)
- medical, dental, and hospital expenses not covered by TRICARE
- funeral expenses
- travel required because of emergency leave or travel and convalescent leave
- payment of rent or a mortgage
- payment of utility bills
- essential repair and/or maintenance of a personally owned vehicle (POV)
- fire or other disaster.

Food voucher assistance is available as a last resort. The Soldier or family member must give a specific reason he or she cannot obtain an AER loan.

Family Advocacy Program

The Family Advocacy Program (FAP) is a specialized prevention and education program that focuses on child abuse, spouse abuse, victim advocacy, and sexual assault. FAP provides the following services:

Crisis Intervention—handles family issues related to abuse, divorce, stress, deployment, single parenting, etc.

First Steps New Parent Support Program—provides parents with information to help them in coping with demands of parenthood and Army life, to increase their parenting knowledge and skills, to enhance the lives of newborns and all children, and to reduce the incidence of child abuse and neglect. Each parent receives a tote bag filled with educational material, including a First Steps Calendar with developmental milestones, a nursery-rhyme CD that is medically proven to effectively soothe newborns, and many other useful items.

Victim Advocacy Program (VAP)—provides services to ensure that any person who is a victim of domestic violence receives support and assistance. The mission is to protect the victims of abuse and treat all family members affected by abuse or involved in it. Services include safety planning, victim advocacy, assisting the victim with court and legal rights, providing education to military and civilian communities, Referral Services Support groups for victims of domestic violence, and emergency shelter.

Sexual Assault Prevention and Response Program (SAPR)—aims to eliminate incidents of sexual assault through education, prevention, integrated victim support, rapid reporting, thorough investigation, appropriate action, and follow-up. Under the Defense Department's confidentiality policy, military sexual assault victims are offered two reporting options: restricted reporting and unrestricted reporting (see Section 1, Values and Ethics Track).

Respite Care—offers temporary child care to relieve parental stress and provide a nurturing and developmentally appropriate environment for children. FAP advocates will evaluate each case and speak directly with the family to determine its needs. Families may receive respite care when attending parenting classes, individual/couples counseling, support groups, or for a stress break due to a family crisis, deployment, or reunion. There is no charge for this service.

Parental Stress Relief Program—is designed to provide stress relief to spouses of deployed Soldiers. The Family Advocacy Program (FAP) coordinates with Child, Youth, and School Services (CYSS) to provide respite child care once a month to promote positive healthy families during deployment. Not only does this program provide respite for the parent left behind, but also quality time for children to play and learn together.

Marriage Retreats—are sponsored by FAP and the Chaplains' Family Life Center for Soldiers and families as part of the reunion process when Soldiers return from deployment.

Employment Readiness Program

The Employment Readiness Program (ERP) assists spouses and teenage children of active and retired service members with opportunities to seek employment on- and off-post. A listing of employment opportunities is updated weekly; these listings include civil service, nonappropriated fund, contract, and local-business opportunities. The program makes job referrals after determining a client's needs and interests through a job-strategy workshop and counseling. It provides additional workshops on a variety of topics, including résumé writing, interviewing techniques, and self-assessment.

Army Family Team Building

Army Family Team Building (AFTB) provides training and information to military personnel, spouses, family members, and civilians in support of the military lifestyle. Strong families are the support pillars behind strong Soldiers. The AFTB mission is to provide the knowledge, skills, and behaviors designed to prepare military families in leadership and self-sufficiency. Training includes an explanation of military terms, acronyms, customs and courtesies, military chain of command and chain of concern, community resources, and much more.

Army Family Action Plan

The Army Family Action Plan (AFAP) relays ideas and suggestions from Army people to the Army leadership. This process allows Soldiers and families to say what is working and what is not—and what they think will make things better. AFAP suggestions alert commanders and Army leaders to areas of concern that need their attention, giving Army leadership the opportunity to move rapidly toward resolving issues.

The AFAP is dedicated to improving standards of living for all military personnel: active duty, Reserve and National Guard Soldiers, spouses, children, retirees, widows and widowers, and all DoD employees.

CONCLUSION

Financial issues will always be a subject of discussion and concern in both a Soldier's home and unit. As a new Army leader, you need to understand Army pay and financial services, not only for the benefit of you and your family, but also for that of your Soldiers and their families.

Key Words

garnishee (garnish)
Servicemembers' Group Life Insurance (SGLI)
Thrift Savings Plan (TSP)
Basic Allowance for Housing (BAH)
Basic Allowance for Subsistence (BAS)
Defense Finance and Accounting Service (DFAS)

Learning Assessment

1. Explain the purpose of the Basic Allowance for Housing (BAH) and when a Soldier receives it.

2. Explain when Soldiers receive the Basic Allowance for Subsistence (BAS).

3. Explain the purpose of the Family Separation Allowance.

4. Explain how a Soldier can obtain a copy of his or her leave and earnings statement.

5. Explain the purpose of managing your financial debt.

6. Compare and contrast an IRA and the TSP.

References

Defense Finance and Accounting Service website. (n.d.). Retrieved 21 October 2008 from http://www.dfas.mil

Field Manual 6-22, *Army Leadership: Competent, Confident, and Agile.* 12 October 2006.

Military Pay: Thrift Savings Plan (TSP). (n.d.). Retrieved 21 October 2008 from http://www.dfas.mil/militarypay/thriftsavingsplantsp.html

STP 14-44A-OFS, *Soldiers Manual and Trainer's Guide AOC 44A: Officer Foundation Standards, Finance Corps (44) Company Grade Officer's Manual, Ranks 2LT, 1LT and CPT.* 17 October 2003.

TC 21-7, *Personal Financial Readiness and Deployability Handbook.* 14 August 2003.

Officership Track

Section

8

INSTALLATION SUPPORT SERVICES FOR SOLDIERS AND DEPENDENTS

Key Points

1 Helping Your Soldiers

2 Army and Defense Department Programs

3 Individual Providers of Support Services Within the Army

4 Civilian Support for Soldiers

The morale of the Soldier is the greatest single factor in war.

Field Marshal Sir Bernard Law Montgomery

Introduction

The US Army depends on the strong **morale** of every Soldier serving. As a platoon leader, you have an important role in building and maintaining morale among your Soldiers. The purpose of this chapter is to describe installation support services, which are resources that the Army and others offer to help ensure a high level of morale among Soldiers and their families. The chapter also provides information on civilian agencies that provide support to Soldiers and their dependents. The more familiar you are with these resources and how to use them appropriately, the more you will be able to help your Soldiers and their families, and the better the morale of your Soldiers will be.

> **morale**
>
> *the state of the spirits of a person or group as exhibited by confidence, cheerfulness, discipline, and willingness to perform assigned tasks*

Clara Barton, Founder of the American Red Cross

Arriving at the northern edge of the infamous "Cornfield" at about noon [17 September 1862], Clara Barton watched as harried surgeons dressed the soldiers' wounds with cornhusks. Army medical supplies were far behind the fast-moving troops at Antietam Battlefield. Miss Barton handed over to grateful surgeons a wagonload of bandages and other medical supplies that she had personally collected over the past year.

Then Miss Barton got down to work. As bullets whizzed overhead and artillery boomed in the distance, Miss Barton cradled the heads of suffering soldiers, prepared food for them in a local farmhouse, and brought water to the wounded men.

As she knelt down to give one man a drink, she felt her sleeve quiver. She looked down, noticed a bullet hole in her sleeve, and then discovered that the bullet had killed the man she was helping.

Undaunted, the unlikely figure in her bonnet, red bow, and dark skirt moved on—and on, and on, and on. Working nonstop until dark, Miss Barton comforted the men and assisted the surgeons with their work.

When night fell, the surgeons were stymied again—this time by lack of light. But Miss Barton produced some lanterns from her wagon of supplies, and the thankful doctors went back to work.

Miss Barton's timely arrival at the battlefield was no easy task. Only the day before, her wagon was mired near the back of the army's massive supply line. Prodded by Miss Barton, her teamsters drove the mules all night to get closer to the front of the line.

Within a few days after the battle, the Confederates had retreated and wagons of extra medical supplies were rolling into Sharpsburg. Miss Barton collapsed from lack of sleep and a budding case of typhoid fever. She returned to Washington lying in a wagon, exhausted and delirious. She soon regained her strength and returned to the battlefields of the Civil War.

[The American Red Cross, which Miss Barton later founded, is one of the key Soldier support services for the US military. The American Red Cross provides services such as emergency communications, financial assistance, and counseling, through 900 chapters in the United States and on 74 military installations around the world.]

National Park Service, Clara Barton—Angel of the Battlefield

Helping Your Soldiers

As a platoon leader, you will need to help your Soldiers deal with personal problems that may affect their ability to carry out their mission. Some of these problems are predictable; others will arise unexpectedly. Some will be emergencies that require immediate action; others will be resolved through long-term intervention. Many of the problems will involve not only the Soldier but also his or her family.

Whatever problems your Soldiers face, you must be ready to help them deal with the problems successfully. Ignoring or denying a problem will only make it worse. Helping your Soldiers cope with their problems is essential for many reasons:

- You are entrusted with the reputation and well-being of the Army
- Dealing with Soldiers' personal issues and nurturing a healthy fighting force is a part of being an Army leader
- Assisting your Soldiers builds trust, productivity, and unit cohesion.

Providing assistance to a Soldier in need is not an easy job, and more often than not you will not have the professional qualifications or training required to properly assist your Soldiers or their dependents. Their problems will often require the intervention of a trained professional. Recognizing this, the Army offers a number of programs to help Soldiers with individual and family problems. These services are part of the Army's focus on and support for the family.

Army and Defense Department Programs

The Army provides Soldier support services during both peacetime and war at all levels. These range from services through the Department of Defense (DoD) and national and community support organizations to those provided at the unit level—and, in the case of the unit chaplain, at the individual level. Once you become familiar with the services that each of these individuals and groups offer, you can help your Soldiers find the most appropriate source of support for their particular needs. You can find more information about the services and programs at typical Army installations by visiting *www.myarmy lifetoo.com*.

Army Community Service (ACS)

Army Community Service (ACS) is a blend of quality-of-life programs in support of Department of Defense activities. Customer-focused and business-based, it offers programs intended to provide support services, education, and information to assist the military, retiree, civilian, and family-member population of all military services. The mission of the ACS is to assist military personnel and their families in every way possible, either by direct assistance or by referral to proper channels.

ACS programs include:

- *Sexual Assault Prevention and Response (SAPR) Program*—SAPR aims to educate and train Soldiers to prevent sexual assault and responds to ensure victims receive sensitive, confidential, and immediate comprehensive care and treatment.

- *Exceptional Family Member Program (EFMP)*—EFMP works with other military and civilian agencies to provide comprehensive and coordinated medical, educational, housing, community-support, and personnel services to families with special needs.

- *Mobilization and Deployment Assistance (MOB/Deployment)*—MOB/Deployment offers assistance to Soldiers and families before, during, and after deployment.

- *Family Advocacy Program (FAP)*—FAP is dedicated to the prevention, prompt reporting, intervention, and treatment of spouse and child abuse, as well as enhancement of the quality of life for military and family members.

- *Personal Financial Readiness Program (PFRP)*—The aim of PFRP is to educate Soldiers and their families in money management, proper use of credit, financial planning, and their rights as consumers to prevent financial difficulties.

- *Army Emergency Relief (AER)*—AER provides emergency financial assistance to Soldiers, family members, and retirees who are in crisis.

- *Employment Readiness Program (ERP)*—ERP provides information and referral services in the areas of employment, education, training, transition, and volunteer opportunities to give family members the competitive edge needed to secure employment.

- *Information and Referral Program (I&R)*—I&R provides Soldiers and their families with any information and assistance they may need concerning available support programs or will refer them to the appropriate agency for assistance.

- *Relocation Assistance Program (RELO)*—RELO provides services necessary to support Department of Army Personnel and their families as they relocate.

- *Army Volunteer Corps (AVC)*—AVC coordinates volunteers and volunteer opportunities available on the installation.

- *Army Family Team Building Program (AFTB)*—AFTB provides family members and Soldiers valuable information and teaches techniques on how to become more knowledgeable about the military lifestyle and how to develop leadership skills.

- *Army Family Action Plan Program (AFAP)*—AFAP provides a way for Soldiers and family members to let the Army know what works, what doesn't, and propose solutions.

As ACS demonstrates, the Army brings out the best in America's spirit of community involvement and volunteerism. Soldiers and their families know that the Army, other Army spouses, and family members in their community care about them and stand ready to help.

Army Education Center (AEC)

The AEC assists with Veterans Administration and Army education programs, distance learning and continuing education programs, and technical training. AEC also provides individualized learning services and can help Soldiers secure grants or loans to defray college expenses. As a platoon leader, you should become very familiar with your installation or unit education center so you can encourage all your Soldiers to take advantage of the various programs that will enable them to further their education, obtain certifications and degrees, or earn their associate's, bachelor's, or advanced degree. All these educational activities will aid them in gaining advancement in the Army.

DoDEA Schools

The Department of Defense Education Activity (DoDEA), a civilian agency of the Defense Department, operates more than 200 schools for the children of military service members and DoD civilian employees in 13 foreign countries, seven states, Guam, and Puerto Rico. The DoDEA instructional program provides a comprehensive pre-kindergarten through twelfth grade curriculum that can compete with that of any school system in the United States. DoDEA maintains a high school graduation rate of approximately 97 percent. Just as in any school, students who excel scholastically and athletically are eligible for scholarships and grants. In the continental United States, DoDEA schools are located in Alabama, Georgia, Kentucky, New York, North Carolina, South Carolina, and Virginia.

Equal Opportunity Office and Equal Employment Opportunity Office

The mission of the Equal Opportunity Office and Equal Employment Opportunity Office is to ensure equitable treatment and opportunity for all personnel. The Equal Opportunity Office and the Equal Employment Opportunity Office offer assistance in matters involving discrimination due to race, color, national origin, gender, and religion. They also provide information on procedures for initiating complaints and for resolving complaints informally. The missions of the two offices are similar, but they serve different people: The EO office supports Soldiers, while the EEO office supports DA civilians.

Family Readiness Center

The Family Readiness Center helps families handle difficult situations during a Soldier's deployment. Its mission is to develop a state of well-being by building strong military families. The FRC offers training programs on subjects such as preparing for deployment, coping with separation, financial planning, stress management, and preparing for reuniting the family after deployment. When Soldiers deploy, the FRC becomes the life hub that keeps the dependents connected to their deployed Soldiers. The FRC will hold routine meetings to keep dependents informed of the unit's missions and successes.

Children play at the Family Readiness Center child-care room.

Housing Referral Office

The Housing Referral Office (HRO) helps military families and single Soldiers find suitable housing on or off the base. Each time you undergo a permanent change of station (PCS), your commander will authorize you up to 10 calendar days of permissive temporary duty (PTDY) to use toward finding your dependents suitable housing if housing is not immediately available on post. Ten days is not a lot of time to find an apartment or house to rent or home to buy. The HRO can help you narrow your search when finding a house or apartment to rent. The office has a list of pre-screened apartment complexes or landlords who meet credibility and reliability standards when renting to Soldiers. This does not mean you may never have a landlord-tenant issue. However, selecting a place from the HRO list reduces the chances of getting a bad landlord or substandard apartment or house. The HRO will assist in resolving landlord-tenant disputes and can also refer your case to Legal Services if you need legal representation. HRO investigations can result in a landlord's removal from the HRO preferred renters list. If you purchase your home and wish to rent it out to carefully screened military personnel, you can request that HRO list your home for rent after you move. You can have the peace of mind that you are providing a home for a fellow Soldier while having the confidence that you will have fewer problems collecting rent from a Soldier.

Installation Safety Office

Most installations have safety offices staffed by professional safety specialists. These offices provide safety training and information for a whole range of Army activities, including information on accident prevention in sports, recreational activities, and motor vehicle use. For example, the safety office provides programs for motorcyclists that Soldiers must take and pass to ride a motorcycle on post.

Legal Services (SJA)

The Office of the Staff Judge Advocate is the main resource for legal services for all Army personnel. It covers the complete scope of legal activities from criminal prosecution and defense through civil and administrative areas. Specific offices under the SJA include:

The *Legal Assistance Office (LAO)* assists Soldiers with their personal legal affairs by meeting their needs for information on legal matters and resolving legal problems whenever possible. Soldiers and emergency-essential Army civilian employees must be prepared for immediate mobilization and deployment. This means that their personal legal affairs must be in order at all times. The LAO provides information or assistance on matters such as contracts, citizenship, adoption, marital problems, taxes, wills, powers of attorney, bills of sale, domestic relations (adoption, separation, nonsupport), change of name, notarizations, civil rights, depositions, immigration, passports, and damage to personal property. LAO refers clients to civilian lawyers when appropriate.

The *JAG Tax Assistance Center* provides tax assistance and electronic filing to military personnel, retirees, and their family members.

The *Claims Division* adjudicates claims from Soldiers and their family members for damage to or loss of property during government shipment or storage, and for certain other instances of damage to or loss of property. The division also adjudicates claims for personal injury or damage to property caused by the negligent acts or omissions of government personnel. In addition, the division pursues affirmative claims against household-goods carriers for damage to Soldiers' property and against third parties who damaged government property or who caused injuries to Soldiers resulting in medical care at government expense.

The *Administrative and Civil Law Division* provides legal support to commanders and staff in a variety of areas. These include civilian personnel and labor law, contract and fiscal law, environmental law, federal litigation, the Freedom of Information and Privacy Acts, the Joint Ethics Regulation and federal ethics laws, military administrative law (such as line of duty investigations), reports of survey, and military personnel law.

The *Criminal Law Division* advises commanders and staff concerning all adverse actions against Soldiers, including nonjudicial punishment (Article 15), administrative separations, administrative reprimands, and courts-martial. In addition, the division also advises the Military Police and Criminal Investigation Command (CID) concerning criminal investigations. In addition, the division is responsible for prosecuting courts-martial and US Magistrate Court cases.

The *US Army Trial Defense Service* is an independent agency responsible for providing defense counsel to Soldiers facing criminal investigation or disciplinary action, including administrative separation, nonjudicial punishment (Article 15), and court-martial.

Family and Morale, Welfare, and Recreation Command (FMWRC)

The Army's Family and MWR Command is a network of support and leisure services designed to enhance the lives of Soldiers (active, Reserve, and Guard), their families, civilian employees, military retirees, and other eligible participants. Programs include sports, recreation, entertainment, travel and lodging, competitive sports, outdoor recreational programs, special events, family support services (ACS), and Child, Youth, and School Services. MWR contributes to the Army's strength and readiness by offering services that reduce stress, build skills and self-confidence, and foster *esprit de corps*. The Army MWR website is *www. armymwr.com*. Army MWR services include but are not limited to the following:

Armed Forces Recreation Center (AFRC) resorts are affordable joint service facilities operated by the US Army Community and Family Support Center and located at ideal vacation destinations, such as Germany, Hawaii, Florida, Korea, and Virginia seaboards. AFRCs offer a full range of resort hotel opportunities for service members, their families, and other members of the Total Defense Force. AFRCs are self-supporting and funded by nonappropriated fund revenues generated internally from operations. Revenues from AFRCs are continually reinvested to maintain and improve the physical plant while providing the greatest possible value for AFRC guests. Providing high-quality, affordable resort-style facilities at the AFRCs is commensurate with the Army Chief of Staff's philosophy that Soldiers are entitled to the same quality of life as the citizens they are pledged to defend. The Army continues to promote strong family values by providing the AFRCs—a reflection of its strong commitment to improved quality of life. AFRC room rates are affordable and based on rank, pay grade, duty status, room size, and/or room location. Soldiers may also use recreation centers operated by any of the other military services.

Child, Youth, and School Services (CYSS) is an integral part of the Army's support services mission. Under the slogan of "Availability, Affordability, and Quality," CYSS supports Soldiers and their families with hundreds of child-care facilities and thousands of family child-care homes worldwide. Assessed and accredited by outside professional organizations, its child-care program has been called a "model for the nation."

An Army Soldier's workday often includes irregular or extended duty hours, deployments, and temporary duty assignments. A Soldier often has no choice when called for duty. The challenges for care continue to grow as mission requirements demand more of the modern-day Soldier. Child, Youth, and School Services responds to these requirements by providing programs when Soldiers and families most need them—early morning, late evening, weekends, and 24 hours a day, if necessary.

CYSS reduces the conflict between mission readiness and parental responsibility. With CYSS support, Soldiers can concentrate on their missions knowing that their children and young people are safe and supervised by trained and professional staff while participating in quality developmental programs. You can find more information on CYSS at *www.armymwr.com.*

Army Community Recreation resources in the United States and abroad provide mission-sustaining fitness, recreation, and library programs, such as community centers; aquatics activities; gyms and fitness facilities for cardiovascular, strength, and flexibility activities; competitive sports; and information, reference, and research. Community support programs also include outdoor recreation programs and equipment, travel and registration information, Better Opportunities for Single Soldiers, arts and crafts, automotive crafts, performing arts, entertainment, recreational swimming, and recreation. As with the recreation center resorts described above, Soldiers and their families may visit facilities of the other military services. Below are some examples of recreation facilities at specific locations:

- Fort Wainwright—Birch Hill Ski and Snowboard Area
- West Point (US Military Academy)—Skiing
- Portsmouth Naval Shipyard—Ski Tickets
- Annapolis Naval Station—Carr Creek Marina
- Fort Lewis—Northwest Adventure Center
- Aberdeen Proving Ground—Equipment Resource Center
- Carlisle Barracks—Recreational Equipment Rental
- Brunswick Naval Air Station—Outdoor Adventure Programs.

Army libraries are a worldwide network of over 125 libraries supporting Army Knowledge requirements. Wherever Soldiers are stationed—Germany, Bosnia, Kuwait, Korea—the Army general library program will be there. General libraries have something for everyone—from story hours to book discussion groups, from Internet access to word processing, from the latest novels to "how-to" books, from encyclopedias and other reference books to online reference resources. Librarians are available to assist you in searches for information.

The *Better Opportunities for Single Soldiers (BOSS) Program* is designed to serve as the collective voice of single Soldiers through the chain of command. It provides a formal structure through which single Soldiers can speak out on issues regarding their well-being. It provides Soldiers with opportunities to participate in activities that improve their lives, increase self-sufficiency, and develop leadership skills. BOSS also provides Soldiers with opportunities for community service. The program is based on the premise that programs that promote well-being such as this have a positive impact on readiness and retention of a quality force.

If your Soldiers are not participating in BOSS activities, you may want to encourage them to do so. Participation in BOSS can help boost a single Soldier's morale. The more single Soldiers in a unit who are involved in the recreational and community-based activities supported by BOSS, the greater the likelihood of strong overall group morale.

BOSS sponsors activities such as cookouts, laser tag, go-cart racing, movies, and golf or bowling outings. It also helps organize complete vacation trips for single Soldiers.

Army Lodging Facilities, available at many installations, welcome all official and unofficial travelers at rates below most motels and hotels. Soldiers and their families may stay at other military services' lodging facilities at most locations.

Army Craft Centers and *Automotive Skills Centers* provide facilities, tools, equipment, and instructions to Soldiers in woodworking, metalworking, and other hobbies. The Automotive Skills Centers provide professional level maintenance facilities and equipment for maintenance and repair of personal motor vehicles on US Army installations worldwide.

Army and Air Force Exchange Service (AAFES)

The Army and Air Force Exchange Service (AAFES) is a joint military activity providing quality merchandise and services to active duty, Guard, and Reserve members; military retirees; and their families at competitively low prices. AAFES returns earnings to the Army and Air Force to improve troops' quality of life and to provide a dividend to support MWR programs. AAFES operates more than 3,100 facilities worldwide, in more than 30 countries, five US territories, and 49 states. AAFES operates some 160 retail stores and 2,008 fast food restaurants, such as Taco Bell, Burger King, Popeyes, and Cinnabon. AAFES also provides military communities with convenience, specialty stores, and movie theaters on installations worldwide, including locations in Operations Enduring Freedom and Iraqi Freedom. AAFES is also a major source of employment for members of the Army and Air Force family. Approximately 25 percent of the more than 48,000 AAFES associates are military family members.

Individual Providers of Support Services Within the Army

Adjutant General (AG)

The Adjutant General provides support to individuals and families across a broad scope of administrative functions—similar to those of a human resources executive in the civilian world, but much more comprehensive. These duties include providing assistance with personnel and administrative matters, such as orders, identification cards, retirement assistance, deferments, dependent enrollment into the Defense Eligibility and Enrollment Reporting System (DEERS), and in- or out-processing. The installation AG also runs the Casualty Assistance program and will provide survivor-benefit briefings to surviving spouses and assist the spouses in getting the benefits they deserve.

The *AG Transition Office* provides help and information on employment and other important transition issues—such as post-discharge continuing medical care—to Army personnel and their families upon separation from the Army. Transition offices may assign an individual counselor for a Soldier to aid with transition issues. The Transition Office will provide the Soldier with retirement, job-search, and VA benefit briefings.

Chaplain

On 29 July 1775, the Continental Congress established the Chaplain Corps as part of the US Army. Chaplains provide spiritual and humanitarian counseling to Soldiers and their families. The influence of chaplains on the morale of the unit and in spiritual matters has been recognized throughout the Army's history. Today's commander recognizes the value of the chaplain, both in combat and during peacetime.

Army chaplains, in providing religious services and ministries to the command, are instruments of the US government. They help ensure that Soldiers can exercise their freedom of religion guaranteed under the First Amendment of the US Constitution. Although chaplains are ecclesiastically endorsed representatives of their respective faiths, they are trained to avoid favoring any particular religion.

Generally the Soldier seeks a chaplain's services, but chaplains may take a proactive role, as well. In fact, a chaplain has the responsibility to confront the commander if he or she believes that the religious rights of any Soldier are being violated. The unit or installation chaplain can be instrumental in helping the commander connect Soldiers or dependents with the installation support services they need. Chaplains often work hand in hand with many ACS and Family Advocacy programs to ensure Soldiers' and dependents' needs are met. In addition, during times of war, the unit chaplain can be a physical presence for your Soldiers that provides them comfort and consolation as well as assisting them with their spiritual needs as necessary.

Health Services

The Army provides comprehensive health care to Soldiers and their families, from emergency room services to child immunizations to mental health and wellness programs. Your Soldiers and their family members receive free or government-subsidized medical and dental care. The United States Army Health Services Command has established US Army Medical Centers (MEDCEN), US Army Medical Department Activities (MEDDAC), and US Army Dental Activities (DENTAC) at all major Army installations and activities to help with medical and dental services. These services are furnished to eligible beneficiaries—including Soldiers and their families—under an area support concept.

In many cases, your Soldiers and their families will be treated by Army medical and dental personnel, though sometimes they will be treated by contract medical personnel on the installation or by private providers off post. Army health care falls under an overall program known as TRICARE. Active Duty personnel are automatically enrolled in TRICARE. When military medical facilities are not available, civilian inpatient and outpatient care is provided for Soldiers and their families.

Army medical and dental services are among the best in the world, and the TRICARE system is highly responsive, but health-care needs can be complex and confusing. As a leader, you must closely monitor the care your Soldiers and their families receive to make sure their needs and concerns are met. Nothing else is more important to their morale and well-being.

Community Mental Health Service (CMHS). In time of war, both Soldiers and their families back home fear for one another's safety. Single parents also face new responsibilities and challenges. During peacetime, Soldiers and their dependents face the same problems as do other families; frequent moves and the uncertainty of military life may compound these problems.

The CMHS provides various types of counseling and psychiatric evaluations to Soldiers and their family members. Either a commander or a physician can refer to CMHS Soldiers who communicate or demonstrate behavior normally associated with stress or mental health issues. A command referral is a recordable event that appears in the Soldier's records and should be used only as a last-resort intervention method. Prior to sending a Soldier to see the commander, you should make all possible attempts to get the Soldier to make a self-referral to CMHS. As a platoon leader, you must recognize the early warning signs and encourage your Soldiers to get help or, if necessary, take them to get help. If all your attempts fail, you should refer the troubled or stressed Soldier to your commander.

Family Advocacy Officer

The Family Advocacy Officer coordinates support programs for Soldiers' families at military installations and in the surrounding communities. These services include counseling, educational programs, and investigations of child and spousal abuse. As a platoon leader, you may have the unfortunate and unpleasant opportunity to be interviewed or called as a witness to alleged domestic abuse committed by a Soldier, a dependent, or both. In most cases, these types of investigations cross many lines and jurisdictions, involving local, state, and military authorities. Off-post authorities are required to report incidents to the installation's Family Advocacy Officer. It is also important to know that the recommendation the case review committee (CRC) makes to your commander is independent of any civil or criminal charges your Soldier or the dependent may face off post. The most important thing to remember as a platoon leader when dealing with the Family Advocacy Officer or Social Work Office is the Soldier's and dependent's right to confidentiality. Nothing will destroy a Soldier's trust in his or her leaders faster than a confidentiality breach about something as personal and embarrassing as a Family Advocacy or Social Work Office investigation.

Inspector General (IG)

The Army IG is an extension of the eyes, ears, voice, and conscience of the commander. He or she can be the commander's sounding board for sensitive issues within the command. The IG investigates in order to identify injustices affecting Soldiers and their families and eliminates conditions determined to be detrimental to the efficiency, economy, morale, and reputation of the Army. The IG also investigates fraud, waste, and abuse.

IGs are fact-finders and play the role of the honest broker. IGs do not interpret and have no authority over command policies; rather they investigate to determine if anyone in the command violated established regulations or policies. Upon completion of an investigation, the IG will submit findings and make recommendations to the commander. IGs operate within an environment consisting of you, your Soldiers, family members, Army civilian employees, retirees, and other civilians needing assistance with Army matters and the IG system. IGs must maintain a clear distinction between their roles as extensions of the commanders and their sworn duty to serve as impartial problem solvers. It is important that officers understand both sides of the IG's role.

Civilian Support for Soldiers

American Red Cross (ARC)

Since its founding in 1881, the American Red Cross has been the nation's leading emergency-response organization. The ARC offers support and comfort to members of the military and their families during and after emergencies and personal crises. Among the services provided by the American Red Cross are:

- Emergency family communications
- Response to health and welfare inquiries
- Information, referral, and advocacy
- Humanitarian and hardship reassignment
- Discharge review and correction of veterans' military records
- Emergency financial assistance
- Health and safety courses
- Volunteer opportunities.

Through the International Committee of the Red Cross, which offers neutral humanitarian care to the victims of war, the ARC also aids victims of natural disasters. The most important role the American Red Cross plays in the lives of you and your Soldiers is what is known as the Red Cross message system. You must ensure that your Soldiers, their dependents, and their immediate family members all know how to get a Red Cross message to your Soldiers.

In the event of a medical emergency, a natural disaster, or a death in the Soldier's immediate family, the Soldier's spouse or immediate family member can contact the local American Red Cross office and ask that office to contact the American Red Cross office on the Soldier's installation to deliver the emergency message. The installation Red Cross office will contact the installation's 24-hour Emergency Operations Center (EOC), the Installation Operations Center (IOC), or the MP desk, whichever is responsible for 24-hour emergency message delivery on your installation. The 24-hour emergency center will immediately contact the Soldier's 24-hour staff duty officer (SDO), who will in turn immediately contact the Soldier's commander. The American Red Cross logs and tracks every call to ensure the Soldier receives the message within 24 hours of logging the Red Cross message.

The information in the Red Cross message assists the Soldier and his or her commander in determining whether the Soldier should take emergency leave to attend to the family emergency. The Red Cross message also acts as a catalyst for ACS to assist the Soldier with emergency funds and coordinating with airlines, trains, or buses to get the Soldier and his or her dependents a reduced bereavement fare to make emergency travel more affordable for the Soldier. You can find more information about ARC services to Soldiers and their families at *http://www.redcross.org/services/afes/0,1082,0_321_,00.html.*

Army Emergency Relief (AER)

AER is a private nonprofit organization incorporated in 1942 by the secretary of War and the Army chief of staff. Its sole mission is to help Soldiers and their dependents. It can provide emergency financial assistance to Soldiers and their dependents when there is a valid need. These funds may be used for a variety of purposes. AER also provides personal budget counseling and coordinates student loans.

Dedicated to "Helping the Army Take Care of Its Own," AER helps officers fulfill their responsibility for maintaining the morale and welfare of Soldiers. As a platoon leader, you must recognize the early indicators that your Soldiers are having financial problems. Many times poor performance or misconduct is a symptom of a larger problem. You need to identify those Soldiers with financial problems and make appointments for them to take advantage of financial and budgeting classes offered by ACS or AER. If your attempts fail and your Soldier is facing a financial crisis, you need to send your Soldier to see your commander as soon as possible. Your commander can make a command referral for the Soldier to AER and recommend an AER loan or grant to stop the downward spiral of the Soldier's finances. AER can also coordinate for reduced-cost transportation for your Soldier and dependents in order to attend funeral services of an immediate family member. The AER website is *http://www.aerhq.org/.*

United Services Organization (USO)

The USO is a private civilian organization that maintains centers away from military installations for the relaxation and recreation of service members and their families. Many USO centers have snack bars, game rooms, reading rooms, travel and tour centers, and other recreational facilities. Some USO centers are located in major airports and provide sleeping berths for adults and children who are awaiting flights.

The USO works with the Armed Forces to provide live entertainment for installations and hospitals in the United States and overseas. Many USOs offer discount tickets for plays and movies and can recommend places to stay overnight. The USO now includes help for military personnel and family members with social problems, such as drug abuse, family troubles, and financial stress. Its website is located at *www.uso.org.*

Critical Thinking

You are on deployment in Iraq. A Soldier informs you that his family back home has just been left homeless by a hurricane, and he is desperately worried about his wife and children, whom he is now having trouble contacting. What can you do to help him?

Critical Thinking

One of your high-speed team leaders has been late to first formation for two days straight. You also get a call from the on-post daycare center that he is not current for his child's daycare fees. You and your platoon sergeant provide him counseling, and during the counseling session, you learn that he fell behind in his car payments and his creditor has made several attempts to repossess the car. He is hiding the car at a friend's house. How will you assist your team leader? What agencies would you refer your team leader to? Would you leave it up to the team leader to make his own appointments, or would you have your platoon sergeant make the appointments and task the Soldier's squad leader to accompany him on his appointments? Do you try to help the team leader yourself or do you tell your commander about the team leader? How long do you wait before telling your commander about the problems the team leader is having?

Critical Thinking

Your platoon is deployed in Iraq and is at 87 percent strength. You are in the middle of a 12-month tour. Your commander informs you that your newest Soldier, PFC Jones, has a Red Cross message and reads it to you over your radio. The message is from the primary care physician for PFC Jones' mother. The physician is verifying that PFC Jones' mother has terminal cancer and estimates that she has less than a week to live. The physician is requesting PFC Jones' presence on the behalf of the grieving father and siblings. Your company commander wants your recommendation whether to allow her to take emergency leave or keep her on the ground to do her job. Your platoon sergeant reminds you that PFC Jones is new and she took most of her accrued leave to visit her ill mother between graduating in advanced individual training (AIT) and reporting to your unit. She has only four days of accrued leave. You also know that if you recommend sending her back, your commander will want a recommendation on how much advance leave to grant her and will want to know how your platoon will cover your assigned mission when your platoon strength drops to 84 percent. What will you do?

CONCLUSION

This chapter has described several agencies that provide support to Soldiers and their families. By becoming familiar with the services offered by each agency, keeping alert to your Soldiers' needs, and helping them secure appropriate services, you can ensure that your platoon's morale and performance remain high. This is a significant part of your job as a platoon leader—don't neglect it. As an Army leader, you will be placed in positions many times to make the tough calls at your level or to make tough recommendations to your company commander—tough calls that can lift a Soldier's morale sky high or send it diving head-first into the dirt. Support is a two-way street: You support your Soldiers and their families, and your Soldiers will support you.

Key Word

morale

Learning Assessment

1. Explain why it is important to help your Soldiers and their families.
2. Describe the responsibility of a platoon leader in directing Soldiers and their families to service support that meets their needs.
3. Identify some of the agencies in the Army and civilian sector that provide support service to Soldiers.

References

Army OneSource. (n.d.). Retrieved 30 October 2008 from http://aos.myarmylifetoo.com/skins/aos/display.aspx?action=display_page&mode=User&ModuleID=8cde2e88-3052-448c-893d-d0b4b14b31c4&ObjectID=bc0a34d0-dc77-42d1-a3ab-48a94a7b0b4a

DA PAM 385-5, *Fundamentals of Safety in Army Sports and Recreation*. 15 November 1981.

DA PAM 608-42, *Handbook on Information and Referral Service for Army Community Service Centers*. 1 August 1985.

DA PAM 608-47, *A Guide to Establishing Family Support Groups*. 16 August 1993.

Field Manual 6-22, *Army Leadership: Competent, Confident, and Agile*. 12 October 1999.

Fort Campbell KY: Home of the 101st Airborne. (n.d.). Fort Campbell, Ky. Retrieved 30 October 2008 from http://www.campbell.army.mil/

Fort Campbell MWR. (n.d.). Fort Campbell, Ky. Retrieved 30 October 2008 from http://www.fortcampbellmwr.com

Fort Lee, Virginia: Morale, Welfare and Recreation. (30 October 2008). Fort Lee, VA. Retrieved 30 October 2008 from http://www.leemwr.com

National Park Service. (7 December 2001). *Clara Barton—Angel of the Battlefield*. Retrieved 5 December 2005 from http://www.nps.gov/anti/clara.htm

INTRODUCTION TO THE COMBAT LIFESAVER COURSE

Key Points

1 Tactical Combat Casualty Care

2 Performing Care Under Fire

3 Performing Tactical Field Care

4 Performing Combat Casualty Evacuation Care

Leadership in the field depends to an important extent on one's legs, and stomach, and nervous system, and one's ability to withstand hardships, and lack of sleep, and still be disposed energetically and aggressively to command men, to dominate men in the battlefield.

GEN George Marshall

Introduction

The Army battle doctrine was developed for a mobile and widely dispersed battlefield. The doctrine recognizes that battlefield conditions will limit the ability of trained medical personnel, including combat medics, to provide immediate, far-forward care. Therefore, the Army developed a plan to provide additional care to injured combat Soldiers. The Combat Lifesaver Course is part of that plan. You will be required to attend and graduate from the course and certify as a Combat Lifesaver, adding to your leadership skills. A Combat Lifesaver (CLS) must recertify every year or attend the entire course again. This section will introduce you to some of the procedures you will need to know to certify as a CLS.

It's important to recognize that up to 90 percent of combat deaths occur on the battlefield before the casualties reach a medical treatment facility (MTF). Most of these deaths are inevitable (massive trauma, massive head injuries, and so forth). Some conditions, however—such as bleeding from a wound in an arm or leg, **tension pneumothorax**, and airway problems—can be treated on the battlefield. This treatment can be the difference between a Soldier dying on the battlefield or recovering in an MTF. Estimates hold that proper use of self-aid, buddy-aid, and Combat Lifesaver skills can reduce battlefield deaths by 15 percent.

But remember that in combat, functioning as a CLS is your *secondary mission*. Your combat duties remain your primary mission. Your first priority while under fire is to return fire and kill the enemy. You should render care to injured Soldiers only when such care does not endanger your primary mission.

> **tension pneumothorax**
>
> *lung collapse*

TABLE 1.1	Estimated Breakdown of Battlefield Deaths
DEATHS DUE TO GROUND COMBAT	
31 percent—Penetrating head trauma	
25 percent—Surgically uncorrectable torso trauma	
10 percent—Potentially correctable surgical trauma	
9 percent—Exsanguination (bleeding out) from extremity wounds	
7 percent—Mutilating blast trauma	
5 percent—Tension pneumothorax	
1 percent—Airway problems	
12 percent—Died of wounds after evacuation to a MTF, mostly from infections and complications of shock	

CLS Course Trains Soldiers to Save Lives

CAMP ATTERBURY, Ind.—When Soldiers are wounded in combat, the most immediate medical care available might be given by other Soldiers on the battlefield, most of whom are not combat medics.

Many Soldiers training at Camp Atterbury, Indiana, are taking the Combat Lifesaver (CLS) Course offered by the 205th Infantry Brigade in order to prepare them for such situations.

"I'm helping a Soldier save a life," said SGT Stacey N. Edwards, CLS instructor, 205th Infantry Brigade. "In Iraq and Afghanistan, 9 out of 10 times, a combat lifesaver will be able to treat a wounded Soldier before a medic will. Skills we are teaching them here save lives over there."

The Army requires 20 percent of personnel in each unit to be Combat Lifesaver certified, said SGT Clint Higgins, CLS instructor, 205th Infantry Brigade.

The four-day course involves 40 hours of training from Soldiers who are combat medics. Eight hours of the course are taught in the warrior training course at Forward Operating Base Bayonet (on Camp Atterbury), said SGT Chris W. Rhea, CLS instructor, 205th Infantry Brigade.

The three main areas of preventable combat deaths addressed in the course are bleeding out, lung collapse, and airway blockage. Nearly 90 percent of these deaths are due to these types of wounds, said SGT Anthony Bussing, a CLS instructor, 205th Infantry Brigade.

If the CLS-certified Soldiers can initially treat these wounds, it will help medics save more lives on the battlefield by enabling the wounded Soldiers to stay alive until a medic can treat them. Not only does the CLS course here teach Soldiers these skills, it also adds the stresses of a combat environment into the training, added Bussing.

In the final exercise of the course, Soldiers break into teams and practice CLS skills on each other in a simulated combat environment. Soldiers must wear their Individual Body Armor, and those Soldiers who are mobilizing must also carry their weapons with them.

In addition to the IBA and weapons, the exercise also incorporates old uniforms for "casualties" to wear in order to make the training as realistic as possible.

"We've been donated old uniforms," said Rhea. "It adds to the realism because the students have to expose the (simulated) wound by cutting through the clothing."

After completing the course, Soldiers become more confident in their ability to keep their fellow Soldiers alive if they are hurt on the battlefield.

"If I have to perform the CLS tasks, I know what needs to be done," said SSG Gregory Dumas Jr., human resources specialist, 2nd Battalion 337th Infantry Regiment (Training Support Battalion), 205th Infantry Brigade. "I won't be so nervous because the hands-on training was very realistic."

During the final exercise, the students also practiced inserting IVs in each other to get hands-on experience.

"I know that I have seen improvement in my ability to initialize an IV since yesterday when we practiced it in the classroom," said Dumas.

Edwards, who is a qualified combat medic, [says requiring] students to apply the skills they were taught in the classroom within a combat environment helps to ensure that the Soldiers will be able to use them effectively anytime a real situation does arise.

Through the CLS course at Camp Atterbury and others like it throughout the Army, Soldiers become better prepared to overcome the stresses and complications, that occur on the battlefield, said Edwards.

The Atterbury Crier

> More than 2,500 Soldiers died in Vietnam because of hemorrhage (bleeding) from extremity wounds (wounds to the arm or leg) even though the Soldiers had no other serious injuries.

care under fire

care rendered when you are under hostile fire and are very limited in the care you can provide

tactical field care

care rendered when you and the casualty are safe, and you are free to provide casualty care to the best of your ability

combat casualty evacuation care

care rendered during casualty evacuation (CASEVAC) aboard nonmedical vehicles or aircraft

> Casualty evacuation (CASEVAC) is different from medical evacuation (MEDEVAC). While CASEVAC uses nonmedical transport, MEDEVAC uses medical vehicles (ground ambulances), medical helicopters (air ambulances), or other medical transportation.

Tactical Combat Casualty Care

There are three types of tactical combat casualty care:

1. **Care under fire:** You are under hostile fire and are very limited in the care you can provide.

2. **Tactical field care:** You and the casualty are safe, and you are free to provide casualty care to the best of your ability.

3. **Combat casualty evacuation care:** You render the care during casualty evacuation (CASEVAC). *Casualty evacuation* means moving casualties aboard nonmedical vehicles or aircraft. You give combat casualty evacuation care while the casualty is awaiting pickup or is being transported by a nonmedical vehicle.

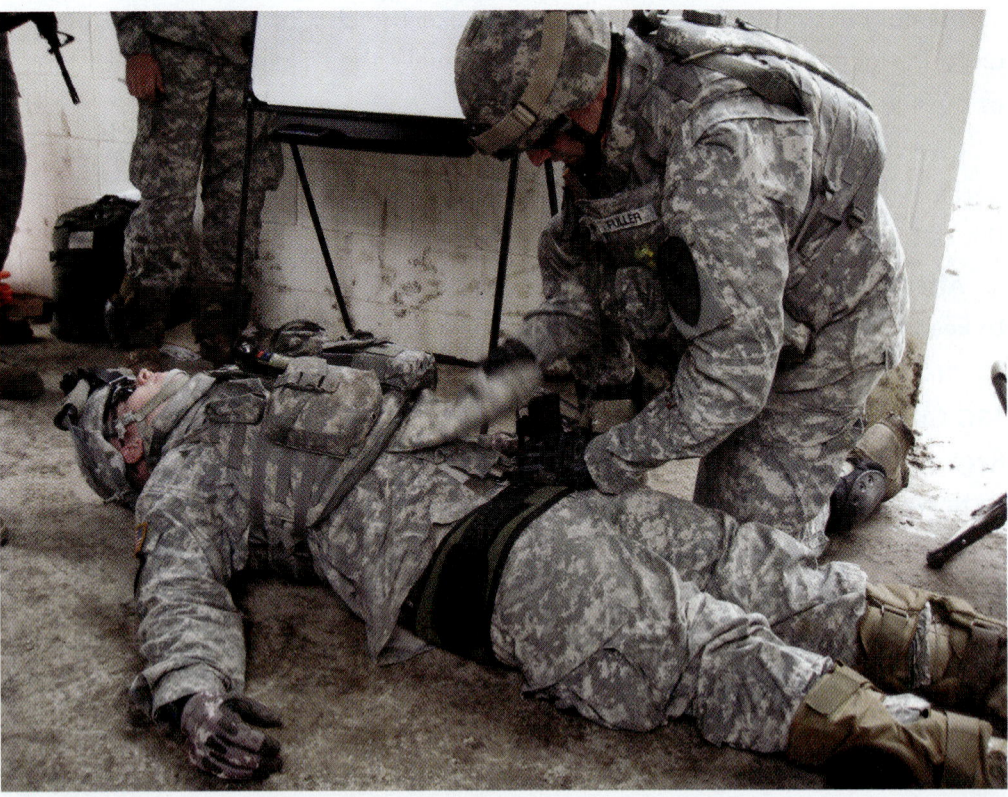

Performing Care Under Fire

You may have to render care at the scene of the injury while you (the CLS) and the casualty are still under effective hostile fire. In such a situation, you should perform the following.

1. Return fire as directed or required before providing medical treatment.

2. Determine if the casualty is alive or dead.

3. Provide tactical care to the live casualty. Remember, though, that reducing or eliminating enemy fire may be more important to the casualty's survival than the treatment you can provide.

   - Suppress enemy fire
   - Use cover or concealment (smoke)
   - Direct the casualty to return fire, move to cover, and administer self-aid (such as taking measures to stop bleeding), if possible—if the casualty can't move, and you can't move the casualty to cover, and the casualty is still under direct enemy fire, have the casualty "play dead"
   - Keep the casualty from sustaining additional wounds
   - Reassure the casualty.

If the casualty has equipment that is essential to the mission, move the mission-essential equipment also. **Do not try to move equipment that is not mission essential.**

You must determine the relative threat of the tactical situation versus the risk to the casualty. Can you remove the casualty to a place of relative safety without becoming a casualty yourself? Is the casualty safer where he or she is? If possible, Soldiers should seek assistance from their leader.

4. If you decide you can safely move the casualty to a safe area, you may need to administer lifesaving care (such as a tourniquet to stop bleeding) before moving the casualty.

   - If the casualty is unresponsive, move the casualty and his or her weapon to cover as the tactical situation permits

   - If the casualty has severe bleeding from a limb (arm or leg) or has suffered amputation of a limb, administer lifesaving hemorrhage control (i.e., apply a tourniquet) before moving the casualty.

5. Soldiers should communicate the medical situation to their team leader.

6. Tactically transport the casualty, his or her weapon, and any mission-essential equipment to cover.

7. Recheck the bleeding-control measures as the tactical situation permits.

Performing Tactical Field Care

You perform tactical field care when you and the casualty are no longer under direct enemy fire. In general, you and your Soldiers should follow the following steps:

1. *Communication.* Soldiers should communicate the medical situation to their unit leader:

   - upon determining that the casualty can't continue the mission

   - before initiating any medical procedures (ensure that the tactical situation allows for time to treat the casualty before initiating any medical procedures)

   - upon any significant change in the casualty's status.

2. *General impression.* Form a general impression of the casualty as you approach (extent of injuries, chance of survival, and so forth). Evaluate the tactical situation.

3. *Level of consciousness.* When possible, determine the casualty's level of consciousness using the **AVPU system**. Ask questions that require more than a "yes" or "no" answer; for example, "What is your name?" "What is the date?" "Where are we?" Recheck the casualty's level of consciousness about every 15 minutes to determine if the casualty's condition has changed.

4. *Airway.* Assess and secure the casualty's airway.

   - If the casualty is conscious, can speak, and is not having trouble breathing, no airway intervention is needed

   - If the casualty is unconscious, perform the following:

     - Use a head-tilt/chin-lift or jaw-thrust to open the airway. The head-tilt/chin-lift method is the normal method of opening the casualty's airway. Use the jaw thrust method if you suspect that the casualty has suffered a spinal injury, in which case you don't want to move the head

You render tactical field care when no longer under effective hostile fire. Tactical field care also applies to situations in which an injury has occurred on a mission, but there is no hostile fire, and the only available medical equipment is what the CLS and individual Soldiers have carried into the field.

Maintaining a check on the casualty's level of consciousness is especially important when the casualty has suffered a head injury.

AVPU system

a system to communicate a casualty's status
A—The casualty is alert, knows who he or she is, the date, where you both are, and so forth
V—The casualty is not alert, but responds to verbal commands
P—The casualty responds to pain, but not to verbal commands
U—The casualty is unresponsive (unconscious)

- Check the casualty for breathing. Place your ear over the casualty's mouth and nose with your face toward the casualty's chest while maintaining the casualty's airway (head-tilt/chin-lift or jaw-thrust). Look for the rise and fall of the casualty's chest and abdomen. Listen for sounds of breathing. Feel for his or her breath on the side of your face. If the casualty is not breathing, begin rescue breathing
- If the casualty is breathing on his or her own, use a nasopharyngeal airway (NPA) device to maintain the airway
- If the casualty has no additional injuries, roll the casualty into the recovery position (on his or her side). This allows accumulated blood and mucus to drain from the casualty's mouth instead of choking the casualty.

5. *Chest.* Assess and treat the casualty for chest injuries.

- Expose the chest and check for equal rise and fall. Remove the minimum of clothing required to expose and treat injuries. Protect the casualty from the environment (heat and cold) as much as possible
- Examine the chest for wounds. Check for both entrance and exit wounds (sucking chest wounds)
- Immediately seal any penetrating injuries to the chest with airtight material. Seal one open chest wound with a three-sided seal (i.e., with one side of airtight material left untaped). Sealing the wound keeps air from entering the wound. If air can freely enter through the wound, the casualty's lung may collapse. The three-sided seal prevents air from entering the chest, but allows trapped air to escape
- Monitor the casualty for increased difficulty breathing (severe respiratory distress—breathing becomes more labored and faster). If respiration becomes progressively worse, assume tension pneumothorax exists and decompress the affected chest side with a 14-gauge needle inserted at the second intercostal space (ICS) on mid-clavicular line (MCL). Secure the catheter in place with tape
- If the casualty has been treated for an open chest wound, position or transport the casualty with the affected (injured) side down, if possible. This way, the body pressure acts to "splint" the affected side.

6. *Bleeding.* Identify and control major bleeding.
   - Apply a tourniquet to a major amputation of the extremity (arm or leg)
   - Apply an emergency trauma bandage and direct pressure to a severely bleeding wound.
     - If conventional methods of controlling severe bleeding (emergency trauma bandage, direct pressure, pressure dressing, hemostatic dressing, and so forth) do not control the bleeding on an extremity, apply a tourniquet
     - If a tourniquet was previously applied, consider changing the tourniquet to a pressure dressing and/or using a hemostatic dressing to control bleeding. Loosen the tourniquet, but do not remove it, while applying conventional methods of controlling bleeding. If conventional methods can't control the hemorrhage, retighten the tourniquet until the bleeding stops.

7. *Intravenous fluids.* Determine if the casualty requires fluid resuscitation (replacement). Use your initial assessment, the casualty's radial pulse (the pulse on the side of the wrist), and the casualty's mental status to determine if fluid resuscitation is required. You can determine these items of information even in the typical noisy and chaotic battlefield environment.
   - If the casualty has only superficial wounds, intravenous (IV) resuscitation is not necessary, but you should encourage the casualty to drink. More than 50 percent of casualties are in this category
   - If the casualty has a significant wound to an extremity or to the trunk (neck, chest, abdomen, or pelvis), the casualty is coherent, and you can feel a radial pulse, insert a saline lock. Do not give intravenous fluids at this time, but continue to monitor the casualty. Begin administering fluids intravenously if the casualty's mental status (which you rate using the AVPU system) decreases, or you can no longer determine his or her radial pulse
   - If the casualty does not have a radial pulse, ensure that the bleeding has been controlled (using direct pressure, pressure dressings, hemostatic bandage, or a tourniquet as needed). Insert a saline lock and begin administering intravenous fluids (one IV bag contains 500 milliliters [ml] of Hextend®) as rapidly as possible. Recheck the casualty's pulse in 30 minutes.
     - If the radial pulse has returned, do not give any additional fluids. Monitor the casualty's pulse as frequently as possible
     - If the radial pulse does not return, give an additional 500 ml of Hextend® and evacuate the casualty as soon as possible.

8. *Other wounds.* Identify and treat other wounds. Dress all wounds, including exit wounds. Remember to remove only the minimum of clothing required to expose and treat the injuries. Protect the casualty against the environment (hot and cold temperatures).

9. *Fractures.* Splint any obvious long bone fractures.

10. *Combat pill pack.* Give pain medications and antibiotics to any Soldier wounded in combat, using the Soldier's combat pill pack. Do not administer your own pack, since you may need it yourself, and you have no extra combat pill packs in your aid bag.

11. *Field Medical Card.* Initiate a DD Form 1380, US Field Medical Card (FMC), to document the casualty's injuries and the treatment given.

Performing Combat Casualty Evacuation Care

In combat casualty evacuation care, you prepare the casualty for evacuation, if needed. If the casualty will be evacuated by medical transport, you may need to prepare a MEDEVAC request. If medical evacuation is not available, prepare the casualty for evacuation using nonmedical means (CASEVAC).

If the casualty can't walk, transport him or her using a SKED® or improvised litter. If you evacuate an unconscious casualty on a nonmedical vehicle, the CLS may need to accompany the casualty to monitor his or her airway, breathing, bleeding, and IV, and to reinforce the casualty's dressings as needed.

TABLE 1.2	The Combat Lifesaver Course

The Combat Lifesaver Course includes the following topics:

Lesson 1: Performing Tactical Combat Casualty Care

Lesson 2: Evaluating a Casualty

Lesson 3: Opening and Managing a Casualty's Airway

Lesson 4: Treating Penetrating Chest Trauma and Decompressing a Tension Pneumothorax

Lesson 5: Controlling Bleeding

Lesson 6: Initiating a Saline Lock and Intravenous Infusion

Lesson 7: Initiating a Field Medical Card

Lesson 8: Requesting a Medical Evacuation (MEDEVAC)

Lesson 9: Evacuating a Casualty Using a SKED® or Improvised Litter

CONCLUSION

This introduction to the Combat Lifesaver Course aims to assist you in the future when you will be required to attend and graduate from the course and certify as a Combat Lifesaver, adding to your leadership skills. Remember, a CLS must recertify every year or attend the entire course again. The CLS program has shown its value on the battlefield since its inception and will continue to assist Soldiers in saving lives.

Key Words

tension pneumothorax
care under fire
tactical field care
combat casualty evacuation care
AVPU system

Learning Assessment

1. List the top three causes of deaths due to ground combat.

2. Compare and contrast the three phases of tactical combat casualty care.

3. Discuss the three basic steps you should take if you encounter a casualty while under fire.

4. Outline the 11 steps to performing tactical field care successfully.

5. Explain when it is important to accompany a casualty you are evacuating on a nonmedical vehicle.

References

AR 350-1, *Army Training and Education*. 13 January 2006.

Campbell, A. (n.d.). CLS course trains Soldiers to save lives. *The Atterbury Crier*. Retrieved 4 November 2008 from http://www.campatterbury.in.ng.mil/cls_trains.htm

DA PAM 350-59, *Army Correspondence Course Program Catalog*. 1 October 2005.

Field Manual 8-10-15, *Employment of the Field and General Hospitals Tactics, Techniques, and Procedures*. 26 March 1997.

Sub-course ISO873, *Combat Lifesaver Instructor Guide*. September 2006.

FORCE PROTECTION IN THE COE AND OPERATIONAL SECURITY

Key Points

1 **What Is Force Protection?**

2 **Planning for Force Protection in the COE**

3 **Force Protection and Terrorism**

In the Global War on Terrorism, we face an insidious and adaptive adversary capable of gathering open source information on our operations and intentions. Do not provide him assistance through uncontrolled release of information that may compromise our own force protection. We are an Army at war and our Soldiers deserve the best operations security (OPSEC) we can provide.

GEN Peter J. Schoomaker

J. A. Capobianco, OPSEC is everyone's responsibility: Changing a mindset, *Infantry Magazine*

Introduction

The Global War on Terrorism has drawn American Soldiers into a new kind of combat against a nontraditional enemy in many parts of the world. But as US forces wage and win those battles in the Contemporary Operating Environment (COE), security and **force protection (FP)** have gained new meaning. A decade ago, force protection measures were something a platoon leader worried about on a linear battlefield. In the nonlinear COE, however, where terrorists or insurgents can strike at any time and from any direction—using a variety of tactics, techniques, and procedures (TTPs)— you must think about force protection constantly.

This new kind of war demands a heightened sense of personal awareness from all officers and enlisted Soldiers to protect the force. This awareness should grow out of your education, training, and preparation and help you respond appropriately to the diverse and often unpredictable challenges you will face. In the following vignette, CPT Daniel Morgan shares some of what he learned about FP while deployed in Iraq.

Force Protection in Iraq

Force protection must remain on the forefront of every leader's mind [while deployed in the COE]. Protecting your Soldiers requires a tough balance between the safety of your Soldiers and mission necessity. Many times in this environment leaders will avoid missions in order to protect Soldiers. This bad habit is not force protection. We protect Soldiers to maintain combat power for mission accomplishment and to bring them home. . . . [T]wo areas demand specific attention—vehicle preparation and compound security.

Vehicle preparation prior to arrival in theater saves lives. As the first combat unit to assume mission in Mosul, we had to learn the hard way. Vehicles must be prepared in a manner that protects the Soldiers from shrapnel and rifle/machine gunfire. A tough decision must be made with respect to sandbags in the trucks. The M998 HMMWV will experience thousands of miles. The weight of a combat-loaded infantry squad with over 50 sandbags will deteriorate a M998 quickly. The sandbags will save the lives of Soldiers, but they do not protect the M998.

Armor plating along the doors of the drivers and passengers and along the benches in the back of the M998 protects Soldiers. On December 26, 2003, we were ambushed while clearing an intersection of IEDs. After one explosion and a fusillade of fire from two enemy machine guns, we inspected the trucks and found that the armor plating on the doors and back of the M998 had withstood the explosion and machine gun impacts, saving the lives of more than 10 Soldiers.

---

force protection (FP)

actions to minimize the effects of enemy firepower, including weapons of mass destruction (WMD), maneuver, and information

---

Leaders do not have the latitude to "avoid" missions. Your job as a leader is to train your Soldiers in the art of force protection on both the operational and personal level (such as shopping, sightseeing, or R&R while stationed in the COE).

The value of a trained warfighter far outweighs the cost of the materiels the warfighter uses. Although a better-armored platform for the warfighter would be best, the time required to prepare these platforms was not available to meet mission timelines— therefore the opening months of the Iraqi campaign resulted in a higher casualty rate than is normally acceptable.

Soldiers searching a vehicle

Security is timeless in military operations. During mounted movements in an urban environment, vehicles must have three-dimensional security. Threats can come from anywhere at anytime. Leaders must prepare their vehicles to facilitate 360-degree security. We placed benches inside every HMMWV (high mobility multipurpose wheeled vehicle) and LMTV (light medium tactical vehicle). I do not know if we were the first ones to do this, but we did recognize this early on, due to AAR comments by Soldiers. An RPG will hit you so fast that if Soldiers are not in the proper security position, you may never know the origin of fire. Simple wooden benches so Soldiers can sit back-to-back improve security, increase offensive capabilities, and enable units to gain the initiative quickly.

Static compound security remains ever-present on the battlefield. Commanders need to balance mission requirements with protecting their company command post or battalion TOC [tactical operations center]. Every compound will be on a road so vehicles can gain access. Some locations permit you to shut down all civilian traffic and some areas will not allow this isolation. The difference in successful or "just-surviving" compound security can be the active versus passive measures taken by a unit.

Static security in an urban area requires a presence outside of the walled compound [that the enemy cannot easily identify]. Commanders need to dispatch patrols during varying times, not only to clear IEDs, but to clear unoccupied buildings, search for fighting positions, occupy OPs, etc. Active, aggressive methods to push your security blanket farther out than your walled compound protects your Soldiers, allowing them to rest and plan comfortably. Commanders must implement a combination of active and passive measures to isolate their company compound as much as possible.

American Soldiers are facing men with a cell phone in one hand, an RPG in the other, and ill-conceived hatred in their heart. This enemy is asymmetric in the most unpredictable way. Technology only enhances the Soldiers' capabilities to kill the enemy. . . . In the end, US Soldiers must meet the enemy—specifically terrorists—face-to-face, hand-to-hand and kill them. Company commanders must bring to bear creativity, aggressiveness, and an offensive spirit to take away the enemy's will. In the end, gather information on enemy targets and then narrowly target them with overwhelming combat power.

Throughout this conflict, I discovered that most things taught in Army schools remain valid and worth remembering during my decision-making process. The most important factor that was reinforced to me that applies to everything discussed here is the necessity to conduct combat AARs after every patrol, whether there was contact or not. Second, troop leading procedures are vital, especially conducting a reconnaissance, rehearsals [pre-combat checks and inspections], and building a terrain model, and supervising platoon and leader operation orders and rehearsals. Third, and most important, maintain an offensive spirit always. Look for the enemy to shoot at you, [maneuver on the enemy using all the assets available to you, and] shoot back and kill or capture them. Bold leaders are dangerous and that is what you want in them as they fight this fight.

Infantry Magazine

> Unit commanders cannot become complacent once their security measures are in place. They must understand that the enemy will eventually identify their security measures and devise a way to penetrate them. This forces unit commanders to constantly add and remove layers of their security measures to keep any possible attackers off balance.

Critical Thinking

Which of the following forms of communication is secure from interception or eavesdropping by terrorists or insurgents in the COE?

- A cordless telephone
- A civilian landline telephone
- Civilian e-mail
- Dot-mil (.mil) e-mail
- A cell phone
- An Internet blog
- An Army website that requires a password to access

What Is Force Protection?

Protection refers to preserving a force's fighting potential so the commander can apply as much combat power as possible at the decisive time and place. In Army doctrine, protection has four components: *force protection, field discipline, safety,* and *avoiding fratricide* (friendly fire).

> *Force protection,* the primary component, minimizes the effects of enemy firepower—including weapons of mass destruction (WMD), maneuver, and information.

> *Field discipline* stems losses and protects resources from hostile environments, such as illnesses and environment-borne diseases.

> *Safety* reduces the inherent risk of noncombat deaths and injuries, such as vehicle and aircraft accidents.

> *Avoiding fratricide* minimizes the inadvertent killing or maiming of Soldiers by friendly fire.

Although you will want to apply protection in all you and your platoon do in the COE, this section will focus on force protection. FP in the COE consists of the measures you take to prevent or minimize the effects of insurgent and terrorist attacks on your platoon—whether traveling or patrolling; in the forward operating base (FOB) or in more garrison-like environments, such as rear echelon areas. Force protection *does not* include actions to defeat the enemy or protect against accidents, weather, or disease.

The force protection measures you employ should adequately address your assessment of the threat. In countries like Iraq, applying too many or overly restrictive FP measures can hinder the effectiveness of your mission—resulting in a delay in returning the country to a sense of normalcy—and can prevent the peace process from continuing and expanding. Additionally, excessive force protection measures can limit interaction between your platoon and the local people, reducing your ability to gather intelligence or conduct negotiations during tense situations. One of the enemy's goals is to drive a wedge of mistrust between you and the local population, so an overly protective standing operating procedure (SOP) or policy can actually benefit your enemy. At the same time, it is quite acceptable to create an outer ring of protection by persuading the local people that combined security efforts will benefit both them and your force.

Planning for Force Protection in the COE

Force protection is an important command responsibility at all levels, including your platoon. As a leader, you must assess the various threats to your platoon and take appropriate measures to protect it. The enemy in the COE will employ terrorist tactics, such as bombings, kidnappings, assassinations, ambushes, and raids. You should address FP during any planning you do for any type of operation and revise your plans as necessary while you are executing them.

Again, this does not mean that you stop patrolling or isolate your Soldiers from contact with the local people. You hinder your mission and even increase the risk to your force if you restrain your Soldiers from conducting prudent operations and establishing an active and capable presence in the area.

Measures to Decrease the Threat and Damage

You can take several actions to bolster force protection and reduce your platoon's vulnerability to attack from insurgents and terrorists. The first step is to know your enemy and his abilities. Once you know some of the threats your force will face, you will be able to develop countermeasures to negate them. For your platoon, this may mean combat patrols, conducting a reconnaissance to gather additional information, or learning how to work with local civilians who may reveal bits of information you can use.

At the platoon level, planning the use of personnel and weapons to enhance force protection is fairly simple, but remember to use all the assets and weapons at your disposal. Platoon leaders must also ensure the use of all available combat multipliers, such as air defense, artillery, engineers, psychological operations, and civil affairs assets, and both rotary and fixed-wing air support that may be attached to your unit or in direct or general support of it.

Remember: The most successful insurgent and terrorist attacks in the last few years have used tactics never tried before. Always think—and train your Soldiers to think— "outside the box." Update your risk assessment frequently.

Force protection is directly linked to the risk management process. An FP measure is nothing more than the mitigating action or countermeasure that your platoon must apply to a known or expected threat or risk to minimize, reduce, or negate the threat or risk. For example, the simple act of moving your platoon around in the COE will require careful route planning. First, you should gather information on enemy abilities from your commander and/or the battalion intelligence officer, as well as from your patrolling. If you are moving as directed by your higher headquarters' latest operation order (OPORD), the order should include an update on enemy abilities. Next, you will consider measures to minimize threats, such as IEDs and ambushes along your route of travel. These measures may extend from configuring your vehicles with sandbags to adjusting tactics, such as the distance between your vehicles, your order of march, and which vehicles you equip with machine guns or grenade launchers.

In assessing the risks and developing the appropriate FP measures, you create a force protection plan. Your FP plan should speak about and react to each COE risk or threat and you should design it to negate, mitigate, or reduce the effects of the threat or risk.

Considerations in your FP plan may include:

Sites, accommodations, and defensive positions. Take precautions to protect positions, equipment, and emplacements. These may include obstacles you have constructed and shelters your Soldiers use. Your unit must practice alert procedures and develop drills to occupy defensive positions rapidly. A robust engineer force can provide support in constructing obstacles, fighting position, and other measures to counter enemy threats. Additionally, engineers could help you build mock fighting positions to deceive the enemy as to your actual position. Since insurgents and terrorists sometimes use mortars to attack friendly forces, mock positions might cause them to deplete their mortar caches while at the same time protecting your Soldiers. You can also use camouflage to deceive enemy fighters as to the location of your positions and equipment.

In a constantly fluid battlefield that expands and contracts, however, the engineer assets are often not available to properly ensure that you can implement all FP measures. Therefore, unit leaders must ensure they utilize all the organic and nonorganic assets available to ensure they can maintain the greatest degree of security possible. In the absence of engineer assets, site location is crucial to prevent the enemy's unimpeded movement in and around your forces. You must maintain proper dispersion of troops and resources. You should also develop evacuation and reaction plans that everyone in the platoon can easily understand and effectively apply to react to a threat and neutralize or avoid it.

Traffic control points (TCPs). Your platoon may be required to establish and man traffic control points along main supply routes (MSRs) or along key routes through your area of operation. Use these TCPs to not only restrict traffic for security purposes, but also inspect vehicles to intercept terrorists, insurgents, and suicide bombers as they try to maneuver in or across your area or deliver munitions for use against US and coalition troops. As a minimum protection, you should select locations that offer superior observations and fields of fire and that are easily defensible with an armed overwatch. Your unit should be able to completely stop or disable a vehicle far enough away from the main body of the organization so that if it detonates it will cause minimal damage to personnel, equipment, or your platoon's ability to accomplish its assigned mission.

Access control points (ACPs). ACPs are very similar to TCPs. The difference is that ACPs control access to the interior of a friendly perimeter. Your platoon may be tasked to prepare or man ACPs to your company FOB.

Personnel vulnerabilities. Forces are always vulnerable to security risks from local nationals and expatriate employees and other personnel who may be subject to bribes, threats, or compromise. The current COE is full of examples of terrorists taking advantage of these vulnerabilities, such as the bombing of a dining facility tent in Mosul, Iraq, by a local food vendor, which killed American and Iraqi Soldiers and local employees. Remember that the threat from local criminal elements is constant and a key FP consideration.

Seeing the same local national every day, who has clearance to work within the boundaries of your cantonment area, does not mean that you can take a lesser FP measure during your screening process. A false sense of friendship or normalcy contributes to complacency. The particular individual in Mosul was willing to give up his life to kill Americans—so you must assume that such individuals are willing to be patient until you trust them enough to allow them to bypass your normal security measures. Such laxity is a practice you cannot afford to adopt.

Personal awareness (situational awareness). The single most effective force-protection measure is your Soldiers' individual awareness in all circumstances. They must look for things that are out of place and the patterns that signal an imminent or unfolding attack. This is the thinking behind the Army's "every Soldier is a sensor" initiative, which seeks to make better use of every Soldier's ability to gather and interpret information. Ensure that your Soldiers remain alert, do not become dulled by routine, and maintain appearance and military bearing. The following vignette relates several episodes that demonstrate the importance of alertness and training in defeating would-be attackers.

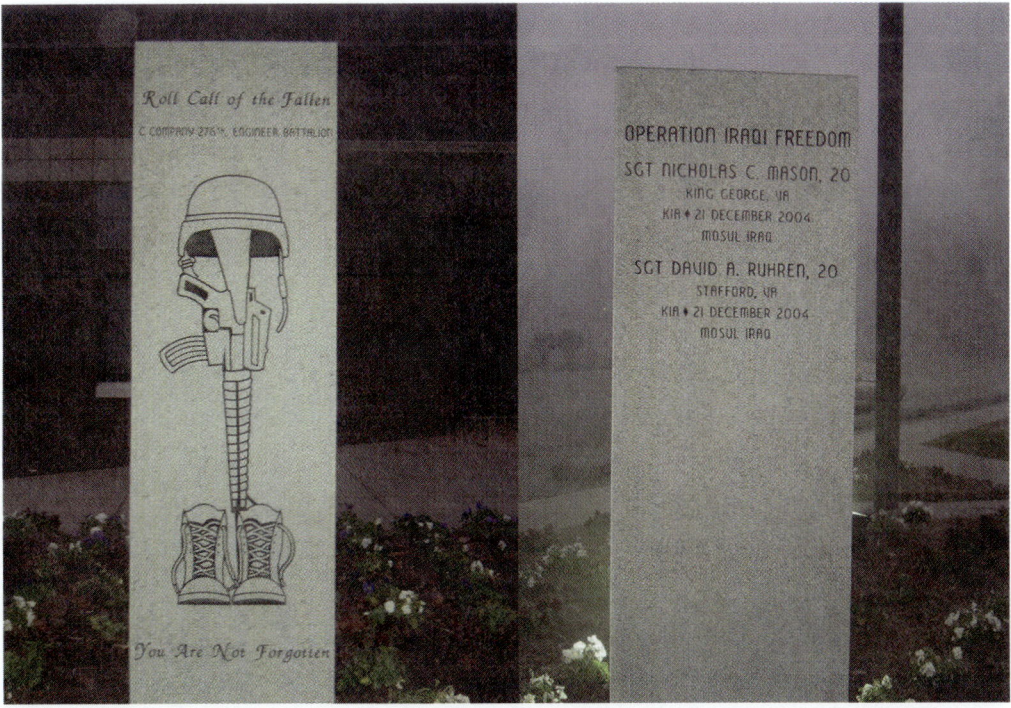

Army SGT Nicholas C. Mason, 20, of King George, Virginia, and Army SGT David A. Ruhren, 20, of Stafford, Virginia, died in Mosul, Iraq, on 21 December 2004 when a suicide bomber entered their dining facility and detonated an improvised explosive device. The Soldiers were assigned to the Army National Guard's 276th Engineer Battalion and to the Army National Guard Armory in West Point, Virginia. One theory is that the suicide bomber may have taken a job as a food vendor in order to gain access to the camp's dining facility.

Soldiers Thwart IED Attacks in Iraq

WASHINGTON—Five large improvised explosive devices on major Baghdad highways were identified by Task Force Baghdad Soldiers Aug. 14 and then safely destroyed before they could hurt anyone.

"We are aggressively and actively countering the Anti-Iraqi Force IED threat by doing more patrols," said Maj. Raul Benitez, intelligence officer for 4th Brigade Combat Team, 3rd Infantry Division. "It is not that the terrorists are doing a poor job of planting these devices, but rather, we are becoming better at identifying these possible threats."

The IEDs were destroyed by an explosive ordnance disposal team.

Training also helped Soldiers from 3/15th Infantry Regiment when their Bradley Fighting Vehicle was struck by an IED Aug. 13 near Sadr City.

Following the attack, the Soldiers exited the vehicle and secured the area. Additional US Army units responded to secure the site while Baghdad civil defense responders worked to extinguish the resulting fire.

Iraqi and US Soldiers squelched a terrorist attack on a patrol base in southwest Baghdad Aug. 11 when terrorists fired five rocket-propelled grenade rounds at the base, along with 10 minutes of rifle fire.

A patrol consisting of Iraqi and US Soldiers set out in the direction of the attack and captured six of the attackers and a weapons cache of two RPGs and three rifles with ammunition hidden nearby.

"We have been making a definite impact in our area of operations," said Lt. Col. Steve McCorkle, commander of the 2nd Battalion, 121st Infantry, 48th Brigade Combat Team.

Iraqi Police also curbed an attack with the arrest of a man in central Baghdad Aug. 11 who was carrying a black bag containing what police thought was TNT. Upon questioning, the man admitted he planned to place the bomb in the area. Task Force Baghdad explosives experts were called to the scene and disposed of the bomb, taking the suspect into custody for further questioning.

Operations conducted by Task Force Baghdad Soldiers Aug. 13 led to the capture of 10 terror suspects and 3 weapons caches. The operations conducted in the pre-dawn hours consisted of cordoning off a number of suspected terrorist safe-houses and searching house by house. Ten suspected terrorists thought to be involved with terrorist activity in south Baghdad were found and detained.

K-9s enhance force protection.

Three weapons caches were also found that day based on tips provided by Iraqi citizens. Two caches were in northwest Baghdad, and one of them consisted of five boxes of anti-aircraft ammunition. The third cache was found in southeast Baghdad and contained RPGs and two launchers and 16 mortar rounds and a launcher.

In northern Iraq, Task Force Freedom detained 21 terrorist suspects in Mosul Aug. 13–14 in several separate operations by Multi-National Forces and Iraqi Security Forces.

Army News Service

Sniper threats. The sniper can pose a significant threat in the COE. Countermeasures include rehearsed responses, reconnaissance and surveillance through aggressive patrolling, and cover and concealment. The **rules of engagement** (ROE) should provide your Soldiers with specific instructions on the amount of force to use in responding to sniper fire. Your platoon SOP or battle drill will also instruct you on how to react. Depending on your location, ROE and SOP may require precision counter-sniper fire to eliminate a sniper and reduce collateral damage. Or your ROE or SOP may allow you to use mortars, tank main gun rounds, artillery, or close air support to take out a sniper.

Security measures. The security measures you implement depend on your METT-TC analysis and your analysis of the risks identified during your risk assessment. The security measures may include the full range of active and passive measures, such as patrolling, reconnaissance and surveillance, the use of quick reaction forces, and the use of other combat multipliers, such as medics, forward observers, scouts, engineers, civil affairs, and interpreters.

Coordination. Whenever possible, you should thoroughly coordinate with nongovernmental organizations (NGOs), local military and civil agencies, local tribal leadership, as well as humanitarian organizations. From them you can gather additional human intelligence (HUMINT) and other information to help you learn of possible attacks, enemy activity, and enemy abilities. Such information might even help you prevent attacks from happening in the first place—one of the best ways to protect your force.

Evacuation. You should have a plan to evacuate your force should conditions warrant it. Whether it's a fire drill to evacuate a building or actions taken under fire, you will minimize casualties if your evacuation is well planned and rehearsed. Your platoon should rehearse the evacuation plan and develop contingency plans, just as you would plan alternate rally points for a raid. More importantly, you must plan to secure your assembly areas or the rally points that you will occupy when reacting to terrorist attacks. It is a common terrorist practice to plan secondary and tertiary explosions in and around known evacuation routes or consolidation areas.

Force Protection Checklist

Since force protection is a primary concern for all leaders, most units should have a published FP plan or SOP. The following checklist may help you enhance your unit's FP measures.

- *Rules of Engagement (ROE).* Do the ROE protect your Soldiers while not alienating the local population?

- *Operational security (OPSEC).* Are your Soldiers disclosing unclassified information that compromises the mission? Is your unit continually evaluating its OPSEC measures?

- ***Physical security (PHYSEC).*** Is access to unit and individual work and billeting areas controlled? Are other safeguards—such as guards, barriers, or patrols—available, if necessary? Do local PHYSEC measures match the terrorist threat condition?

physical security (PHYSEC)

security concerned with physical measures designed to safeguard personnel, to prevent unauthorized access to equipment, installations, materiel, and documents, and to safeguard them against espionage, sabotage, damage, and theft

- *Personnel security (PERSEC)*. Can the unit or individual routines vary, including the time of day certain duties are performed? As far as the mission permits, can individuals blend with the local environment? Are your Soldiers properly or adequately trained in ground combatives (hand-to-hand combat)? Do your Soldiers know the boundaries of your unit's area of operations? Do your Soldiers know the off-limits areas the commander has established?

- *Law enforcement.* Local law enforcement constitutes the unit's outer ring of security and protection. Does liaison with local law enforcement exist? Are law enforcement abilities sufficient to counter the anticipated threat? Are the locations of civilian police, military police, government agencies, the US Embassy, and other safe locations available, and do all Soldiers have that information? Do your Soldiers understand their rights under the Geneva Convention?

- *Antiterrorism.* Is an updated threat briefing available? Does your platoon have a plan to react to or counter a terrorist attack? Has your platoon rehearsed the plan? Does an alert system exist? Can the unit reduce its presence where possible? Is a means in place to identify the location of all personnel at all times?

- *Reaction force.* If all else fails, is your platoon ready to respond? What is your plan if your unit is attacked? What about a plan to assist nearby units and personnel? Do your Soldiers know the criteria that warrant requesting a quick reaction force (QRF)? Do your Soldiers know which unit is tasked as the QRF? If your platoon is the designated QRF, do you have an established plan that addresses priorities of work, and a time standard to respond and react to the QRF alert? Do your Soldiers know the code word used to alert the QRF? Is the code word changed frequently to avoid it being overheard outside your perimeter and therefore compromised? Do you alternate the response routes you take when practicing for the QRF mission?

Critical Thinking

How can Soldiers' alertness help deter a terrorist attack?

Constraints

As with all planning and military operations, you must consider constraints when planning force protection. Much depends on the level of force protection your risk assessment calls for.

The ROE may impose severe limitations, depending on the area in which you operate. It may limit the weapons you can use and the time available for offensive actions. Based on the threat, you can call for additional assets such as personnel, equipment, or indirect fire support.

In addition, your environment may impose constraints. The effects of the terrain may make it difficult to dig fighting positions that protect your Soldiers, or install reinforcing obstacles that restrict or prohibit vehicular traffic. Dispersing your Soldiers improves force protection but may also reduce your ability to mass fires. The higher the FP level, the riskier it is to establish or maintain contact with the local population.

You will have to deal with other limitations, restrictions, or constraints as well, especially local, cultural, and religious considerations that may restrict, limit, or impede employing your own FP measures. As the platoon leader, your problem solving and critical thinking skills may well determine whether your Soldiers survive a terrorist or insurgent attack. Plan and execute wisely.

Orders and Force Protection

Typically, the operation order (OPORD) addresses force protection. For higher headquarters, force protection guidance and instructions are published in various support annexes under "Coordinating instructions." These outline measures to protect units, camps, headquarters, and installations. These instructions should specify which elements need protecting, the unit assigned to provide protection, and any special factors for that protection.

Force Protection and Terrorism

As you have learned, terrorists and insurgents have shown a trend toward an ever-increasing number of attacks and sophistication in their methods. These threats exist both within the COE and around the world.

As noted in the section on the culture of terrorism, terrorist methods include threats, bombing, kidnapping, hostage-taking, hijacking, assassination, sabotage, arson, armed raids or attacks, and other measures to disrupt daily activities. From a military perspective, you must ensure you are implementing FP measures every day and in every location—including your home post.

This protection includes observing and reporting:

- Suspicious surveillance, filming, photography, recording, envelopes, or packages
- Inquiries from unknown persons related to military capabilities, limitations, or operational information
- Suspicious vehicles (cars, trucks, vans) operating or parked in or around the installation
- Phone calls, messages, or e-mails from unknown persons requesting information about deployments, building occupancy, or procedures.

Do not discuss with unknown people or in public places any aspect of military operations or planning; military capabilities or limitations; force-protection measures, capabilities, or posture; or any information related to unit deployments. Train your Soldiers not to do so either.

Antiterrorism and Counterterrorism

Recall the difference between *antiterrorism* and *counterterrorism*. Antiterrorism (AT) is defensive measures used to reduce the vulnerability of individuals and property to terrorist acts, while counterterrorism (CT) refers to offensive measures to prevent, deter, and respond to terrorism. Both are part of the Defense Department (DoD) definition of force protection. CT offensive measures to deter or defeat terrorist attacks can vary from patrols, ambushes, and raids to attacks. Since these types of offensive operations are discussed in other sections, this section will focus on AT measures.

Antiterrorism (AT) Programs

To meet the terrorist threat, every level of command should develop and use a comprehensive AT program that includes a protective posture for units in a peacetime environment. While this applies to such activities as units performing normal duties and serving in security assistance organizations, peacekeeping missions, or mobile training teams, it's only half the battle for units in a hostile environment.

Antiterrorist measures identify and reduce the risk of loss or damage to potential targets. You must develop procedures to detect and deter terrorist activity before it takes place, reducing the probability of a successful terrorist event. Your measures must also include your Soldiers' tactical reaction to an incident, including engaging the terrorists to end the incident with minimal loss of life and property.

The AT program stresses deterring terrorist incidents through preventive measures. Terrorists know the importance of their acts' emotional effect on audiences other than the victims. News media coverage allows terrorists to incite public fear while pursuing their objectives. You can neutralize that exposure by thwarting the activity to begin with.

Soldiers Use Biometrics to Increase Force Protection

FORWARD OPERATING BASE WAR EAGLE, Iraq—With more than 5 million Iraqis and five army and national police brigades in Baghdad, anonymity is a great weapon for insurgents.

Even if a unit identifies suspicious activity by a local national on its base, what is there to stop that individual from moving down the road to a new base? The answer for coalition forces is simple: biometrics.

The Soldiers of Company A, 3rd Special Troops Battalion, 3rd Brigade Combat Team, 4th Infantry Division, Multi-National Division-Baghdad, have fingerprinted, photographed and eye-scanned more than 400 individuals in their efforts to keep Forward Operating Base War Eagle safe.

CPT Mike Poaletti, commander of Headquarters and Headquarters Troop, 3rd BCT, 4th Inf. Div., and the FOB War Eagle mayor, said he appreciates the biometric scanning.

"They enhanced force protection by screening local nationals with the [intelligence] system. Their screening enabled us to keep unwanted personnel off the FOB," he said.

There are two Soldiers who are primarily responsible for the biometric screening and badge issuing operations on FOB War Eagle. SSG Matthew Valek, a Ceres, Calif., native, who serves as the senior non-commissioned officer in charge of badge operations, has seen the benefits of his works. In four months of operations he has identified one previously identified insurgent.

"It is reassuring to know that each local national we hire has not been involved in insurgent activities before. Force protection is important to us," Valek said.

1LT George Rolston, the War Eagle Base Defense Officer, said he agrees.

"Badges are critical. They enable us to quickly identify local nationals and vehicles allowed on the FOB. Their badges explicitly state who is authorized where and with what; it's a big help," said Rolston, a Conrad, Iowa, native, who serves with Company B, 3rd STB, 3rd BCT, 4th Inf. Div.

The Soldiers of Company A and Base Defense work hand in hand to ensure optimal force protection on FOB War Eagle. Their hard work is amplified by other brigade's biometric efforts resulting in a vastly improved, theater-wide tracking effort.

Army.mil/News

Force Protection Conditions

The Department of Defense uses a standardized set of terms to describe the terrorism threat level (TTL) in each country: *low, moderate, significant,* and *high.* The Defense Intelligence Agency (DIA) sets the TTL for each country based on analysis of all available information.

Commanders and leaders at all levels use the TTL with their own threat analyses to develop plans and programs to protect assets within their area of responsibility. These are called *force protection conditions* (FPCON) and are a set of specific security measures the commander uses after considering a variety of factors—including the TTL, current events that increase the risk, observed suspicious activities, and intelligence or other reports.

The graduated series of FPCs range from *Force Protection Conditions Normal* to *Force Protection Conditions Delta.* You will have a process at all levels to raise or lower the FPC, based on assessment of local conditions, specific threat information, and/or guidance from higher headquarters.

Force Protection Conditions ALPHA

ALPHA applies when a general threat of possible terrorist activity exists against personnel and facilities. The nature and extent of the threat is unpredictable, and circumstances do not justify taking Force Protection Conditions BRAVO measures. The measures in this FPC can be maintained indefinitely.

Force Protection Conditions BRAVO

BRAVO applies when an increased and more predictable threat of terrorist activity exists. You must be able to maintain the measures in this FPC for weeks without causing your Soldiers or civilians undue hardship, affecting operational capability, and aggravating relations with local authorities.

Force Protection Conditions CHARLIE

CHARLIE applies when an incident occurs or intelligence is received indicating that some form of terrorist action against personnel and facilities is imminent. Implementing the measures in this FPC for more than a short period will probably create hardship and affect the peacetime activities of the unit and its personnel.

Force Protection Conditions DELTA

DELTA applies in the immediate area where a terrorist attack has occurred or when intelligence indicates that terrorist action against a specific location or person is likely. Normally, this FPC is declared only for a specific location.

Critical Thinking

Reflecting on some of your earlier lessons on squad or platoon tactics, how might you alter your movement formations, movement techniques, and dispersion based on the FPCON level for your area of operation?

CONCLUSION

The Army faces enemies around the globe who will strike at any time and in any place they believe they can succeed. They constantly adapt their tactics to find and take advantage of weaknesses in your force's routine. As a platoon leader, you must become expert in this new kind of asymmetrical warfare, which demands continual vigilance and training.

Force protection in the COE is everyone's job. Letting your guard down for even a moment in the COE is to invite an attack that might lead to your death and the death of your Soldiers. The protection of your Soldiers and equipment in all locations and situations must become a way of life. You accomplish this through situational awareness, constant assessment of the threat, and assurance that all your Soldiers are ready to respond.

Key Words

force protection (FP)
physical security (PHYSEC)
personnel security (PERSEC)

Learning Assessment

1. Define force protection.
2. Explain how FP in the COE is different from traditional warfare.
3. List and explain two detection assets Soldiers use to increase FP.
4. List and describe the four levels of force protection conditions.

References

Brown, D. (17 September 2008). Soldiers use biometrics to increase force protection. *Army.mil/News.* Retrieved 29 October 2008 from http://www.army.mil/-news/2008/09/17/12510-soldiers-use-biometrics-to-increase-force-protection/

Field Manual 3-0, *Operations.* 27 February 2008.

Field Manual 3-07, *Stability Operations and Support Operations.* 20 February 2003.

Field Manual 3-07.31, *Peace Operations.* 26 October 2003.

Field Manual 3-19.30, *Physical Security.* 8 January 2001.

Morgan, D. (2004). Deploying to Iraq? Lessons from an infantry company commander. *Infantry Magazine,* January–February 2004. Retrieved 6 December 2005 from http://www.findarticles.com/p/articles/mi_m0IAV/is_1_93/ai_n6123804

The unanimous Declaration of the thirteen united States of America,

When in the Course of human events, it becomes necessary for one people to dissolve the political bands which have connected them with another, and to assume among the powers of the earth, the separate and equal station to which the Laws of Nature and of Nature's God entitle them, a decent respect to the opinions of mankind requires that they should declare the causes which impel them to the separation.

We hold these truths to be self-evident, that all men are created equal, that they are endowed by their Creator with certain unalienable Rights, that among these are Life, Liberty and the pursuit of Happiness.—That to secure these rights, Governments are instituted among Men, deriving their just powers from the consent of the governed,—That whenever any Form of Government becomes destructive of these ends, it is the Right of the People to alter or to abolish it, and to institute new Government, laying its foundation on such principles and organizing its powers in such form, as to them shall seem most likely to effect their Safety and Happiness. Prudence, indeed, will dictate that Governments long established should not be changed for light and transient causes; and accordingly all experience hath shewn, that mankind are more disposed to suffer, while evils are sufferable, than to right themselves by abolishing the forms to which they are accustomed. But when a long train of abuses and usurpations, pursuing invariably the same Object evinces a design to reduce them under absolute Despotism, it is their right, it is their duty, to throw off such Government, and to provide new Guards for their future security.—Such has been the patient sufferance of these Colonies; and such is now the necessity which constrains them to alter their former Systems of Government. The history of the present King of Great Britain is a history of repeated injuries and usurpations, all having in direct object the establishment of an absolute Tyranny over these States. To prove this, let Facts be submitted to a candid world.

He has refused his Assent to Laws, the most wholesome and necessary for the public good.

He has forbidden his Governors to pass Laws of immediate and pressing importance, unless suspended in their operation till his Assent should be obtained; and when so suspended, he has utterly neglected to attend to them.

He has refused to pass other Laws for the accommodation of large districts of people, unless those people would relinquish the right of Representation in the Legislature, a right inestimable to them and formidable to tyrants only.

He has called together legislative bodies at places unusual, uncomfortable, and distant from the depository of their public Records, for the sole purpose of fatiguing them into compliance with his measures.

He has dissolved Representative Houses repeatedly, for opposing with manly firmness his invasions on the rights of the people.

He has refused for a long time, after such dissolutions, to cause others to be elected; whereby the Legislative powers, incapable of Annihilation, have returned to the People at large for their exercise; the State remaining in the mean time exposed to all the dangers of invasion from without, and convulsions within.

He has endeavoured to prevent the population of these States; for that purpose obstructing the Laws for Naturalization of Foreigners; refusing to pass others to encourage their migrations hither, and raising the conditions of new Appropriations of Lands.

He has obstructed the Administration of Justice, by refusing his Assent to Laws for establishing Judiciary powers.

He has made Judges dependent on his Will alone, for the tenure of their offices, and the amount and payment of their salaries.

He has erected a multitude of New Offices, and sent hither swarms of Officers to harrass our people, and eat out their substance.

He has kept among us, in times of peace, Standing Armies without the Consent of our legislatures.

He has affected to render the Military independent of and superior to the Civil power.

He has combined with others to subject us to a jurisdiction foreign to our constitution, and unacknowledged by our laws; giving his Assent to their Acts of pretended Legislation:

For Quartering large bodies of armed troops among us:

For protecting them, by a mock Trial, from punishment for any Murders which they should commit on the Inhabitants of these States:

For cutting off our Trade with all parts of the world:

For imposing Taxes on us without our Consent:

For depriving us in many cases, of the benefits of Trial by Jury:

For transporting us beyond Seas to be tried for pretended offences:

For abolishing the free System of English Laws in a neighbouring Province, establishing therein an Arbitrary government, and enlarging its Boundaries so as to render it at once an example and fit instrument for introducing the same absolute rule into these Colonies:

For taking away our Charters, abolishing our most valuable Laws, and altering fundamentally the Forms of our Governments:

For suspending our own Legislatures, and declaring themselves invested with power to legislate for us in all cases whatsoever:

He has abdicated Government here, by declaring us out of his Protection and waging War against us.

He has plundered our seas, ravaged our Coasts, burnt our towns, and destroyed the lives of our people.

He is at this time transporting large Armies of foreign Mercenaries to compleat the works of death, desolation and tyranny, already begun with circumstances of Cruelty & perfidy scarcely paralleled in the most barbarous ages, and totally unworthy the Head of a civilized nation.

He has constrained our fellow Citizens taken Captive on the high Seas to bear Arms against their Country, to become the executioners of their friends and Brethren, or to fall themselves by their Hands.

He has excited domestic insurrections amongst us, and has endeavoured to bring on the inhabitants of our frontiers, the merciless Indian Savages, whose known rule of warfare, is an undistinguished destruction of all ages, sexes and conditions.

In every stage of these Oppressions We have Petitioned for Redress in the most humble terms: Our repeated Petitions have been answered only by repeated injury. A Prince whose character is thus marked by every act which may define a Tyrant, is unfit to be the ruler of a free people.

Nor have We been wanting in attentions to our British brethren. We have warned them from time to time of attempts by their legislature to extend an unwarrantable jurisdiction over us. We have reminded them of the circumstances of our emigration and settlement here. We have appealed to their native justice and magnanimity, and we have conjured them by the ties of our common kindred to disavow these usurpations, which, would inevitably interrupt our connections and correspondence. They too have been deaf to the voice of justice and of consanguinity. We must, therefore, acquiesce in the necessity, which denounces our Separation, and hold them, as we hold the rest of mankind, Enemies in War, in Peace Friends.

We, therefore, the Representatives of the United States of America, in General Congress, Assembled, appealing to the Supreme Judge of the world for the rectitude of our intentions, do, in the Name, and by Authority of the good People of these Colonies, solemnly publish and declare, That these United Colonies are, and of Right ought to be Free and Independent States; that they are Absolved from all Allegiance to the British Crown, and that all political connection between them and the State of Great Britain, is and ought to be totally dissolved; and that as Free and Independent States, they have full Power to levy War, conclude Peace, contract Alliances, establish Commerce, and to do all other Acts and Things which Independent States may of right do. And for the support of this Declaration, with a firm reliance on the protection of divine Providence, we mutually pledge to each other our Lives, our Fortunes and our sacred Honor.

The Constitution of the United States

We the People of the United States, in Order to form a more perfect Union, establish Justice, insure domestic Tranquility, provide for the common defense, promote the general Welfare, and secure the Blessings of Liberty to ourselves and our Posterity, do ordain and establish this Constitution for the United States of America.

Article. I.

Section. 1.

All legislative Powers herein granted shall be vested in a Congress of the United States, which shall consist of a Senate and House of Representatives.

Section. 2.

The House of Representatives shall be composed of Members chosen every second Year by the People of the several States, and the Electors in each State shall have the Qualifications requisite for Electors of the most numerous Branch of the State Legislature.

No Person shall be a Representative who shall not have attained to the Age of twenty five Years, and been seven Years a Citizen of the United States, and who shall not, when elected, be an Inhabitant of that State in which he shall be chosen.

The actual Enumeration shall be made within three Years after the first Meeting of the Congress of the United States, and within every subsequent Term of ten Years, in such Manner as they shall by Law direct. The Number of Representatives shall not exceed one for every thirty Thousand, but each State shall have at Least one Representative; and until such enumeration shall be made, the State of New Hampshire shall be entitled to chuse three, Massachusetts eight, Rhode-Island and Providence Plantations one, Connecticut five, New-York six, New Jersey four, Pennsylvania eight, Delaware one, Maryland six, Virginia ten, North Carolina five, South Carolina five, and Georgia three.

When vacancies happen in the Representation from any State, the Executive Authority thereof shall issue Writs of Election to fill such Vacancies.

The House of Representatives shall chuse their Speaker and other Officers; and shall have the sole Power of Impeachment.

Section. 3.

The Senate of the United States shall be composed of two Senators from each State thereof for six Years; and each Senator shall have one Vote.

Immediately after they shall be assembled in Consequence of the first Election, they shall be divided as equally as may be into three Classes. The Seats of the Senators of the first Class shall be vacated at the Expiration of the second Year, of the second Class at the Expiration of the fourth Year, and of the third Class at the Expiration of the sixth Year, so that one third may be chosen every second.

No Person shall be a Senator who shall not have attained to the Age of thirty Years, and been nine Years a Citizen of the United States, and who shall not, when elected, be an Inhabitant of that State for which he shall be chosen.

The Vice President of the United States shall be President of the Senate, but shall have no Vote, unless they be equally divided.

The Senate shall chuse their other Officers, and also a President pro tempore, in the Absence of the Vice President, or when he shall exercise the Office of President of the United States.

The Senate shall have the sole Power to try all Impeachments. When sitting for that Purpose, they shall be on Oath or Affirmation. When the President of the United States is tried, the Chief Justice shall preside: And no Person shall be convicted without the Concurrence of two thirds of the Members present.

Judgment in Cases of Impeachment shall not extend further than to removal from Office, and disqualification to hold and enjoy any Office of honor, Trust or Profit under the United States: but the Party convicted shall nevertheless be liable and subject to Indictment, Trial, Judgment and Punishment, according to Law.

Section. 4.

The Times, Places and Manner of holding Elections for Senators and Representatives, shall be prescribed in each State by the Legislature thereof; but the Congress may at any time by Law make or alter such Regulations, except as to the Places of chusing Senators.

The Congress shall assemble at least once in every Year, and such Meeting shall be on the first Monday in December, unless they shall by Law appoint a different Day.

Section. 5.

Each House shall be the Judge of the Elections, Returns and Qualifications of its own Members, and a Majority of each shall constitute a Quorum to do Business; but a smaller Number may adjourn from day to day, and may be authorized to compel the Attendance of absent Members, in such Manner, and under such Penalties as each House may provide.

Each House may determine the Rules of its Proceedings, punish its Members for disorderly Behaviour, and, with the Concurrence of two thirds, expel a Member.

Each House shall keep a Journal of its Proceedings, and from time to time publish the same, excepting such Parts as may in their Judgment require Secrecy; and the Yeas and Nays of the Members of either House on any question shall, at the Desire of one fifth of those Present, be entered on the Journal.

Neither House, during the Session of Congress, shall, without the Consent of the other, adjourn for more than three days, nor to any other Place than that in which the two Houses shall be sitting.

Section. 6.

The Senators and Representatives shall receive a Compensation for their Services, to be ascertained by Law, and paid out of the Treasury of the United States. They shall in all Cases, except Treason, Felony and Breach of the Peace, be privileged from Arrest during their Attendance at the Session of their respective Houses, and in going to and returning from the same; and for any Speech or Debate in either House, they shall not be questioned in any other Place.

No Senator or Representative shall, during the Time for which he was elected, be appointed to any civil Office under the Authority of the United States, which shall have been created, or the Emoluments whereof shall have been encreased during such time; and no Person holding any Office under the United States, shall be a Member of either House during his Continuance in Office.

Section. 7.

All Bills for raising Revenue shall originate in the House of Representatives; but the Senate may propose or concur with Amendments as on other Bills.

Every Bill which shall have passed the House of Representatives and the Senate, shall, before it become a Law, be presented to the President of the United States: If he approve he shall sign it, but if not he shall return it, with his Objections to that House in which it shall have originated, who shall enter the Objections at large on their Journal, and proceed to reconsider it. If after such Reconsideration two thirds of that House shall agree to pass the Bill, it shall be sent, together with the Objections, to the other House, by which it shall likewise be reconsidered, and if approved by two thirds of that House, it shall become a Law. But in all such Cases the Votes of both Houses shall be determined by yeas and Nays, and the Names of the Persons voting for and against the Bill shall be entered on the Journal of each House respectively.
If any Bill shall not be returned by the President within ten Days (Sundays excepted) after it shall have been presented to him, the Same shall be a Law, in like Manner as if he had signed it, unless the Congress by their Adjournment prevent its Return, in which Case it shall not be a Law.

Every Order, Resolution, or Vote to which the Concurrence of the Senate and House of Representatives may be necessary (except on a question of Adjournment) shall be presented to the President of the United States; and before the Same shall take Effect, shall be approved by him, or being disapproved by him, shall be repassed by two thirds of the Senate and House of Representatives, according to the Rules and Limitations prescribed in the Case of a Bill.

Section. 8.

The Congress shall have Power To lay and collect Taxes, Duties, Imposts and Excises, to pay the Debts and provide for the common Defence and general Welfare of the United States; but all Duties, Imposts and Excises shall be uniform throughout the United States;

To borrow Money on the credit of the United States;

To regulate Commerce with foreign Nations, and among the several States, and with the Indian Tribes;

To establish an uniform Rule of Naturalization, and uniform Laws on the subject of Bankruptcies throughout the United States;

To coin Money, regulate the Value thereof, and of foreign Coin, and fix the Standard of Weights and Measures;

To provide for the Punishment of counterfeiting the Securities and current Coin of the United States;

To establish Post Offices and post Roads;

To promote the Progress of Science and useful Arts, by securing for limited Times to Authors and Inventors the exclusive Right to their respective Writings and Discoveries;

To constitute Tribunals inferior to the supreme Court;

To define and punish Piracies and Felonies committed on the high Seas, and Offences against the Law of Nations;

To declare War, grant Letters of Marque and Reprisal, and make Rules concerning Captures on Land and Water;

To raise and support Armies, but no Appropriation of Money to that Use shall be for a longer Term than two Years;

To provide and maintain a Navy;

To make Rules for the Government and Regulation of the land and naval Forces;

To provide for calling forth the Militia to execute the Laws of the Union, suppress Insurrections and repel Invasions;

To provide for organizing, arming, and disciplining, the Militia, and for governing such Part of them as may be employed in the Service of the United States, reserving to the States respectively, the Appointment of the Officers, and the Authority of training the Militia according to the discipline prescribed by Congress;

To exercise exclusive Legislation in all Cases whatsoever, over such District (not exceeding ten Miles square) as may, by Cession of particular States, and the Acceptance of Congress, become the Seat of the Government of the United States, and to exercise like Authority over all Places purchased by the Consent of the Legislature of the State in which the Same shall be, for the Erection of Forts, Magazines, Arsenals, dock-Yards, and other needful Buildings;—And

To make all Laws which shall be necessary and proper for carrying into Execution the foregoing Powers, and all other Powers vested by this Constitution in the Government of the United States, or in any Department or Officer thereof.

Section. 9.

The Migration or Importation of such Persons as any of the States now existing shall think proper to admit, shall not be prohibited by the Congress prior to the Year one thousand eight hundred and eight, but a Tax or duty may be imposed on such Importation, not exceeding ten dollars for each Person.

The Privilege of the Writ of Habeas Corpus shall not be suspended, unless when in Cases of Rebellion or Invasion the public Safety may require it.

No Bill of Attainder or ex post facto Law shall be passed.

No Capitation, or other direct, Tax shall be laid.

No Tax or Duty shall be laid on Articles exported from any State.

No Preference shall be given by any Regulation of Commerce or Revenue to the Ports of one State over those of another; nor shall Vessels bound to, or from, one State, be obliged to enter, clear, or pay Duties in another.

No Money shall be drawn from the Treasury, but in Consequence of Appropriations made by Law; and a regular Statement and Account of the Receipts and Expenditures of all public Money shall be published from time to time.

No Title of Nobility shall be granted by the United States: And no Person holding any Office of Profit or Trust under them, shall, without the Consent of the Congress, accept of any present, Emolument, Office, or Title, of any kind whatever, from any King, Prince, or foreign State.

Section. 10.

No State shall enter into any Treaty, Alliance, or Confederation; grant Letters of Marque and Reprisal; coin Money; emit Bills of Credit; make any Thing but gold and silver Coin a Tender in Payment of Debts; pass any Bill of Attainder, ex post facto Law, or Law impairing the Obligation of Contracts, or grant any Title of Nobility.

No State shall, without the Consent of the Congress, lay any Imposts or Duties on Imports or Exports, except what may be absolutely necessary for executing it's inspection Laws: and the net Produce of all Duties and Imposts, laid by any State on Imports or Exports, shall be for the Use of the Treasury of the United States; and all such Laws shall be subject to the Revision and Controul of the Congress.

No State shall, without the Consent of Congress, lay any Duty of Tonnage, keep Troops, or Ships of War in time of Peace, enter into any Agreement or Compact with another State, or with a foreign Power, or engage in War, unless actually invaded, or in such imminent Danger as will not admit of delay.

Article. II.

Section. 1.

The executive Power shall be vested in a President of the United States of America. He shall hold his Office during the Term of four Years, and, together with the Vice President, chosen for the same Term, be elected, as follows:

Each State shall appoint, in such Manner as the Legislature thereof may direct, a Number of Electors, equal to the whole Number of Senators and Representatives to which the State may be entitled in the Congress: but no Senator or Representative, or Person holding an Office of Trust or Profit under the United States, shall be appointed an Elector.

The Congress may determine the Time of chusing the Electors, and the Day on which they shall give their Votes; which Day shall be the same throughout the United States.

No Person except a natural born Citizen, or a Citizen of the United States, at the time of the Adoption of this Constitution, shall be eligible to the Office of President; neither shall any Person be eligible to that Office who shall not have attained to the Age of thirty five Years, and been fourteen Years a Resident within the United States.

In Case of the Removal of the President from Office, or of his Death, Resignation, or Inability to discharge the Powers and Duties of the said Office, the Same shall devolve on the Vice President, and the Congress may by Law provide for the Case of Removal, Death, Resignation or Inability, both of the President and Vice President, declaring what Officer shall then act as President, and such Officer shall act accordingly, until the Disability be removed, or a President shall be elected.

The President shall, at stated Times, receive for his Services, a Compensation, which shall neither be increased nor diminished during the Period for which he shall have been elected, and he shall not receive within that Period any other Emolument from the United States, or any of them.

Before he enter on the Execution of his Office, he shall take the following Oath or Affirmation:—"I do solemnly swear (or affirm) that I will faithfully execute the Office of President of the United States, and will to the best of my Ability, preserve, protect and defend the Constitution of the United States."

Section. 2.

The President shall be Commander in Chief of the Army and Navy of the United States, and of the Militia of the several States, when called into the actual Service of the United States; he may require the Opinion, in writing, of the principal Officer in each of the executive Departments, upon any Subject relating to the Duties of their respective Offices, and he shall have Power to grant Reprieves and Pardons for Offences against the United States, except in Cases of Impeachment.

He shall have Power, by and with the Advice and Consent of the Senate, to make Treaties, provided two thirds of the Senators present concur; and he shall nominate, and by and with the Advice and Consent of the Senate, shall appoint Ambassadors, other public Ministers and Consuls, Judges
of the supreme Court, and all other Officers of the United States, whose Appointments are not herein otherwise provided for, and which shall be established by Law: but the Congress may by Law vest the Appointment of such inferior Officers, as they think proper, in the President alone, in the Courts of Law, or in the Heads of Departments.

The President shall have Power to fill up all Vacancies that may happen during the Recess of the Senate, by granting Commissions which shall expire at the End of their next Session.

Section. 3.

He shall from time to time give to the Congress Information of the State of the Union, and recommend to their Consideration such Measures as he shall judge necessary and expedient; he may, on extraordinary Occasions, convene both Houses, or either of them, and in Case of Disagreement between them, with Respect to the Time of Adjournment, he may adjourn them to such Time as he shall think proper; he shall receive Ambassadors and other public Ministers; he shall take Care that the Laws be faithfully executed, and shall Commission all the Officers of the United States.

Section. 4.

The President, Vice President and all civil Officers of the United States, shall be removed from Office on Impeachment for, and Conviction of, Treason, Bribery, or other high Crimes and Misdemeanors.

Article. III.

Section. 1.

The judicial Power of the United States shall be vested in one supreme Court, and in such inferior Courts as the Congress may from time to time ordain and establish. The Judges, both of the supreme and inferior Courts, shall hold their Offices during good Behaviour, and shall, at stated Times, receive for their Services a Compensation, which shall not be diminished during their Continuance in Office.

Section. 2.

The judicial Power shall extend to all Cases, in Law and Equity, arising under this Constitution, the Laws of the United States, and Treaties made, or which shall be made, under their Authority;—to all Cases affecting Ambassadors, other public Ministers and Consuls;—to all Cases of admiralty and maritime Jurisdiction;—to Controversies to which the United States shall be a Party;—to Controversies between two or more States;—between Citizens of different States;—between Citizens of the same State claiming Lands under Grants of different States, and between a State, or the Citizens thereof, and foreign States, Citizens or Subjects.

In all Cases affecting Ambassadors, other public Ministers and Consuls, and those in which a State shall be Party, the supreme Court shall have original Jurisdiction. In all the other Cases before mentioned, the supreme Court shall have appellate Jurisdiction, both as to Law and Fact, with such Exceptions, and under such Regulations as the Congress shall make.

The Trial of all Crimes, except in Cases of Impeachment, shall be by Jury; and such Trial shall be held in the State where the said Crimes shall have been committed; but when not committed within any State, the Trial shall be at such Place or Places as the Congress may by Law have directed.

Section. 3.

Treason against the United States, shall consist only in levying War against them, or in adhering to their Enemies, giving them Aid and Comfort. No Person shall be convicted of Treason unless on the Testimony of two Witnesses to the same overt Act, or on Confession in open Court.

The Congress shall have Power to declare the Punishment of Treason, but no Attainder of Treason shall work Corruption of Blood, or Forfeiture except during the Life of the Person attainted.

Article. IV.

Section. 1.

Full Faith and Credit shall be given in each State to the public Acts, Records, and judicial Proceedings of every other State. And the Congress may by general Laws prescribe the Manner in which such Acts, Records and Proceedings shall be proved, and the Effect thereof.

Section. 2.

The Citizens of each State shall be entitled to all Privileges and Immunities of Citizens in the several States.

A Person charged in any State with Treason, Felony, or other Crime, who shall flee from Justice, and be found in another State, shall on Demand of the executive Authority of the State from which he fled, be delivered up, to be removed to the State having Jurisdiction of the Crime.

Section. 3.

New States may be admitted by the Congress into this Union; but no new State shall be formed or erected within the Jurisdiction of any other State; nor any State be formed by the Junction of two or more States, or Parts of States, without the Consent of the Legislatures of the States concerned as well as of the Congress.

The Congress shall have Power to dispose of and make all needful Rules and Regulations respecting the Territory or other Property belonging to the United States; and nothing in this Constitution shall be so construed as to Prejudice any Claims of the United States, or of any particular State.

Section. 4.

The United States shall guarantee to every State in this Union a Republican Form of Government, and shall protect each of them against Invasion; and on Application of the Legislature, or of the Executive (when the Legislature cannot be convened), against domestic Violence.

Article. V.

The Congress, whenever two thirds of both Houses shall deem it necessary, shall propose Amendments to this Constitution, or, on the Application of the Legislatures of two thirds of the several States, shall call a Convention for proposing Amendments, which, in either Case, shall be valid to all Intents and Purposes, as Part of this Constitution, when ratified by the Legislatures of three fourths of the several States, or by Conventions in three fourths thereof, as the one or the other Mode of Ratification may be proposed by the Congress; Provided that no Amendment which may be made prior to the Year One thousand eight hundred and eight shall in any Manner affect the first and fourth Clauses in the Ninth Section of the first Article; and that no State, without its Consent, shall be deprived of its equal Suffrage in the Senate.

Article. VI.

All Debts contracted and Engagements entered into, before the Adoption of this Constitution, shall be as valid against the United States under this Constitution, as under the Confederation.

This Constitution, and the Laws of the United States which shall be made in Pursuance thereof; and all Treaties made, or which shall be made, under the Authority of the United States, shall be the supreme Law of the Land; and the Judges in every State shall be bound thereby, any Thing in the Constitution or Laws of any State to the Contrary notwithstanding.

The Senators and Representatives before mentioned, and the Members of the several State Legislatures, and all executive and judicial Officers, both of the United States and of the several States, shall be bound by Oath or Affirmation, to support this Constitution; but no religious Test shall ever be required as a Qualification to any Office or public Trust under the United States.

Article. VII.

The Ratification of the Conventions of nine States, shall be sufficient for the Establishment of this Constitution between the States so ratifying the Same.

The Word, "the," being interlined between the seventh and eighth Lines of the first Page, the Word "Thirty" being partly written on an Erazure in the fifteenth Line of the first Page, The Words "is tried" being interlined between the thirty second and thirty third Lines of the first Page and the Word "the" being interlined between the forty third and forty fourth Lines of the second Page.

Attest William Jackson Secretary

Done in Convention by the Unanimous Consent of the States present the Seventeenth Day of September in the Year of our Lord one thousand seven hundred and Eighty seven and of the Independence of the United States of America the Twelfth In witness whereof We have hereunto subscribed our Names,

G. Washington
President and deputy from Virginia

Delaware
Geo: Read
Gunning Bedford jun
John Dickinson
Richard Bassett
Jaco: Broom

Maryland
James McHenry
Dan of St Thos. Jenifer
Danl. Carroll

Virginia
John Blair
James Madison Jr.

North Carolina
Wm. Blount
Richd. Dobbs Spaight
Hu Williamson

South Carolina
J. Rutledge
Charles Cotesworth Pinckney
Charles Pinckney
Pierce Butler

Georgia
William Few
Abr Baldwin

New Hampshire
John Langdon
Nicholas Gilman

Massachusetts
Nathaniel Gorham
Rufus King

Connecticut
Wm. Saml. Johnson
Roger Sherman

New York
Alexander Hamilton

New Jersey
Wil: Livingston
David Brearley
Wm. Paterson
Jona: Dayton

Pennsylvania
B. Franklin
Thomas Mifflin
Robt. Morris
Geo. Clymer
Thos. FitzSimons
Jared Ingersoll
James Wilson
Gouv Morris

The Bill of Rights

AMENDMENT I

Congress shall make no law respecting an establishment of religion, or prohibiting the free exercise thereof; or abridging the freedom of speech, or of the press; or the right of the people peaceably to assemble, and to petition the Government for a redress of grievances.

AMENDMENT II

A well regulated Militia, being necessary to the security of a free State, the right of the people to keep and bear Arms, shall not be infringed.

AMENDMENT III

No Soldier shall, in time of peace be quartered in any house, without the consent of the Owner, nor in time of war, but in a manner to be prescribed by law.

AMENDMENT IV

The right of the people to be secure in their persons, houses, papers, and effects, against unreasonable searches and seizures, shall not be violated, and no Warrants shall issue, but upon probable cause, supported by Oath or affirmation, and particularly describing the place to be searched, and the persons or things to be seized.

AMENDMENT V

No person shall be held to answer for a capital, or otherwise infamous crime, unless on a presentment or indictment of a Grand Jury, except in cases arising in the land or naval forces, or in the Militia, when in actual service in time of War or public danger; nor shall any person be subject for the same offence to be twice put in jeopardy of life or limb; nor shall be compelled in any criminal case to be a witness against himself, nor be deprived of life, liberty, or property, without due process of law; nor shall private property be taken for public use, without just compensation.

AMENDMENT VI

In all criminal prosecutions, the accused shall enjoy the right to a speedy and public trial, by an impartial jury of the State and district wherein the crime shall have been committed, which district shall have been previously ascertained by law, and to be informed of the nature and cause of the accusation; to be confronted with the witnesses against him; to have compulsory process for obtaining witnesses in his favor, and to have the Assistance of Counsel for his defence.

AMENDMENT VII

In Suits at common law, where the value in controversy shall exceed twenty dollars, the right of trial by jury shall be preserved, and no fact tried by a jury, shall be otherwise re-examined in any Court of the United States, than according to the rules of the common law.

AMENDMENT VIII

Excessive bail shall not be required, nor excessive fines imposed, nor cruel and unusual punishments inflicted.

AMENDMENT IX

The enumeration in the Constitution, of certain rights, shall not be construed to deny or disparage others retained by the people.

AMENDMENT X

The powers not delegated to the United States by the Constitution, nor prohibited by it to the States, are reserved to the States respectively, or to the people.

The Constitution: Amendments 11–27

Constitutional Amendments 1–10 make up what is known as The Bill of Rights. *Amendments 11–27 are listed below.*

AMENDMENT XI

Passed by Congress March 4, 1794. Ratified February 7, 1795.

Note: Article III, section 2, of the Constitution was modified by amendment 11.

The Judicial power of the United States shall not be construed to extend to any suit in law or equity, commenced or prosecuted against one of the United States by Citizens of another State, or by Citizens or Subjects of any Foreign State.

AMENDMENT XII

Passed by Congress December 9, 1803. Ratified June 15, 1804.

Note: A portion of Article II, section 1 of the Constitution was superseded by the 12th amendment.

The Electors shall meet in their respective states and vote by ballot for President and Vice-President, one of whom, at least, shall not be an inhabitant of the same state with themselves; they shall name in their ballots the person voted for as President, and in distinct ballots the person voted for as Vice-President, and they shall make distinct lists of all persons voted for as President, and of all persons voted for as Vice-President, and of the number of votes for each, which lists they shall sign and certify, and transmit sealed to the seat of the government of the United States, directed to the President of the Senate;—the President of the Senate shall, in the presence of the Senate and House of Representatives, open all the certificates and the votes shall then be counted; —The person having the greatest number of votes for President, shall be the President, if such number be a majority of the whole number of Electors appointed; and if no person have such majority, then from the persons having the highest numbers not exceeding three on the list of those voted for as President, the House of Representatives shall choose immediately, by ballot, the President. But in choosing the President, the votes shall be taken by states, the representation from each state having one vote; a quorum for this purpose shall consist of a member or members from two-thirds of the states, and a majority of all the states shall be necessary to a choice. [And if the House of Representatives shall not choose a President whenever the right of choice shall devolve upon them, before the fourth day of March next following, then the Vice-President shall act as President, as in case of the death or other constitutional disability of the President.—]* The person having the greatest number of votes as Vice-President, shall be the Vice-President, if such number be a majority of the whole number of Electors appointed, and if no person have a majority, then from the two highest numbers on the list, the Senate shall choose the Vice-President; a quorum for the purpose shall consist of two-thirds of the whole number of Senators, and a majority of the whole number shall be necessary to a choice. But no person constitutionally ineligible to the office of President shall be eligible to that of Vice-President of the United States.

Superseded by section 3 of the 20th amendment.

AMENDMENT XIII

Passed by Congress January 31, 1865. Ratified December 6, 1865.

Note: A portion of Article IV, section 2, of the Constitution was superseded by the 13th amendment.

Section 1.
Neither slavery nor involuntary servitude, except as a punishment for crime whereof the party shall have been duly convicted, shall exist within the United States, or any place subject to their jurisdiction.

Section 2.
Congress shall have power to enforce this article by appropriate legislation.

AMENDMENT XIV

Passed by Congress June 13, 1866. Ratified July 9, 1868.

Note: Article I, section 2, of the Constitution was modified by section 2 of the 14th amendment.

Section 1.
All persons born or naturalized in the United States, and subject to the jurisdiction thereof, are citizens of the United States and of the State wherein they reside. No State shall make or enforce any law which shall abridge the privileges or immunities of citizens of the United States; nor shall any State deprive any person of life, liberty, or property, without due process of law; nor deny to any person within its jurisdiction the equal protection of the laws.

Section 2.
Representatives shall be apportioned among the several States according to their respective numbers, counting the whole number of persons in each State, excluding Indians not taxed. But when the right to vote at any election for the choice of electors for President and Vice-President of the United States, Representatives in Congress, the Executive and Judicial officers of a State, or the members of the Legislature thereof, is denied to any of the male inhabitants of such State, being twenty-one years of age,* and citizens of the United States, or in any way abridged, except for participation in rebellion, or other crime, the basis of

representation therein shall be reduced in the proportion which the number of such male citizens shall bear to the whole number of male citizens twenty-one years of age in such State.

Section 3.

No person shall be a Senator or Representative in Congress, or elector of President and Vice-President, or hold any office, civil or military, under the United States, or under any State, who, having previously taken an oath, as a member of Congress, or as an officer of the United States, or as a member of any State legislature, or as an executive or judicial officer of any State, to support the Constitution of the United States, shall have engaged in insurrection or rebellion against the same, or given aid or comfort to the enemies thereof. But Congress may by a vote of two-thirds of each House, remove such disability.

Section 4.

The validity of the public debt of the United States, authorized by law, including debts incurred for payment of pensions and bounties for services in suppressing insurrection or rebellion, shall not be questioned. But neither the United States nor any State shall assume or pay any debt or obligation incurred in aid of insurrection or rebellion against the United States, or any claim for the loss or emancipation of any slave; but all such debts, obligations and claims shall be held illegal and void.

Section 5.

The Congress shall have the power to enforce, by appropriate legislation, the provisions of this article.

Changed by section 1 of the 26th amendment.

AMENDMENT XV

Passed by Congress February 26, 1869. Ratified February 3, 1870.

Section 1.

The right of citizens of the United States to vote shall not be denied or abridged by the United States or by any State on account of race, color, or previous condition of servitude—

Section 2.

The Congress shall have the power to enforce this article by appropriate legislation.

AMENDMENT XVI

Passed by Congress July 2, 1909. Ratified February 3, 1913.

Note: Article I, section 9, of the Constitution was modified by amendment 16.

The Congress shall have power to lay and collect taxes on incomes, from whatever source derived, without apportionment among the several States, and without regard to any census or enumeration.

AMENDMENT XVII

Passed by Congress May 13, 1912. Ratified April 8, 1913.

Note: Article I, section 3, of the Constitution was modified by the 17th amendment.

The Senate of the United States shall be composed of two Senators from each State, elected by the people thereof, for six years; and each Senator shall have one vote. The electors in each State shall have the qualifications requisite for electors of the most numerous branch of the State legislatures.

When vacancies happen in the representation of any State in the Senate, the executive authority of such State shall issue writs of election to fill such vacancies: *Provided,* That the legislature of any State may empower the executive thereof to make temporary appointments until the people fill the vacancies by election as the legislature may direct.

This amendment shall not be so construed as to affect the election or term of any Senator chosen before it becomes valid as part of the Constitution.

AMENDMENT XVIII

Passed by Congress December 18, 1917. Ratified January 16, 1919. Repealed by amendment 21.

Section 1.

After one year from the ratification of this article the manufacture, sale, or transportation of intoxicating liquors within, the importation thereof into, or the exportation thereof from the United States and all territory subject to the jurisdiction thereof for beverage purposes is hereby prohibited.

Section 2.

The Congress and the several States shall have concurrent power to enforce this article by appropriate legislation.

Section 3.

This article shall be inoperative unless it shall have been ratified as an amendment to the Constitution by the legislatures of the several States, as provided in the Constitution, within seven years from the date of the submission hereof to the States by the Congress.

AMENDMENT XIX

The right of citizens of the United States to vote shall not be denied or abridged by the United States or by any State on account of sex.

Congress shall have power to enforce this article by appropriate legislation.

AMENDMENT XX

Passed by Congress March 2, 1932. Ratified January 23, 1933.

Note: Article I, section 4, of the Constitution was modified by section 2 of this amendment. In addition, a portion of the 12th amendment was superseded by section 3.

Section 1.
The terms of the President and the Vice President shall end at noon on the 20th day of January, and the terms of Senators and Representatives at noon on the 3d day of January, of the years in which such terms would have ended if this article had not been ratified; and the terms of their successors shall then begin.

Section 2.
The Congress shall assemble at least once in every year, and such meeting shall begin at noon on the 3d day of January, unless they shall by law appoint a different day.

Section 3.
If, at the time fixed for the beginning of the term of the President, the President elect shall have died, the Vice President elect shall become President. If a President shall not have been chosen before the time fixed for the beginning of his term, or if the President elect shall have failed to qualify, then the Vice President elect shall act as President until a President shall have qualified; and the Congress may by law provide for the case wherein neither a President elect nor a Vice President shall have qualified, declaring who shall then act as President, or the manner in which one who is to act shall be selected, and such person shall act accordingly until a President or Vice President shall have qualified.

Section 4.
The Congress may by law provide for the case of the death of any of the persons from whom the House of Representatives may choose a President whenever the right of choice shall have devolved upon them, and for the case of the death of any of the persons from whom the Senate may choose a Vice President whenever the right of choice shall have devolved upon them.

Section 5.
Sections 1 and 2 shall take effect on the 15th day of October following the ratification of this article.

Section 6.
This article shall be inoperative unless it shall have been ratified as an amendment to the Constitution by the legislatures of three-fourths of the several States within seven years from the date of its submission.

AMENDMENT XXI

Passed by Congress February 20, 1933. Ratified December 5, 1933.

Section 1.
The eighteenth article of amendment to the Constitution of the United States is hereby repealed.

Section 2.
The transportation or importation into any State, Territory, or Possession of the United States for delivery or use therein of intoxicating liquors, in violation of the laws thereof, is hereby prohibited.

Section 3.
This article shall be inoperative unless it shall have been ratified as an amendment to the Constitution by conventions in the several States, as provided in the Constitution, within seven years from the date of the submission hereof to the States by the Congress.

AMENDMENT XXII

Passed by Congress March 21, 1947. Ratified February 27, 1951.

Section 1.
No person shall be elected to the office of the President more than twice, and no person who has held the office of President, or acted as President, for more than two years of a term to which some other person was elected President shall be elected to the office of President more than once. But this Article shall not apply to any person holding the office of President when this Article was proposed by Congress, and shall not prevent any person who may be holding the office of President, or acting as President, during the term within which this Article becomes operative from holding the office of President or acting as President during the remainder of such term.

Section 2.
This article shall be inoperative unless it shall have been ratified as an amendment to the Constitution by the legislatures of three-fourths of the several States within seven years from the date of its submission to the States by the Congress.

AMENDMENT XXIII

Passed by Congress June 16, 1960. Ratified March 29, 1961.

Section 1.
The District constituting the seat of Government of the United States shall appoint in such manner as Congress may direct:

A number of electors of President and Vice President equal to the whole number of Senators and Representatives in Congress to which the District would be entitled if it were a State, but in no event more than the least populous State; they shall be in addition to those appointed by the States, but they shall be considered, for the purposes of the election of President and Vice President, to be electors appointed by a State; and they shall meet in the District and perform such duties as provided by the twelfth article of amendment.

Section 2.
The Congress shall have power to enforce this article by appropriate legislation.

AMENDMENT XXIV

Passed by Congress August 27, 1962. Ratified January 23, 1964.

Section 1.
The right of citizens of the United States to vote in any primary or other election for President or Vice President, for electors for President or Vice President, or for Senator or Representative in Congress, shall not be denied or abridged by the United States or any State by reason of failure to pay poll tax or other tax.

Section 2.
The Congress shall have power to enforce this article by appropriate legislation.

AMENDMENT XXV

Passed by Congress July 6, 1965. Ratified February 10, 1967.

Note: Article II, section 1, of the Constitution was affected by the 25th amendment.

Section 1.
In case of the removal of the President from office or of his death or resignation, the Vice President shall become President.

Section 2.
Whenever there is a vacancy in the office of the Vice President, the President shall nominate a Vice President who shall take office upon confirmation by a majority vote of both Houses of Congress.

Section 3.
Whenever the President transmits to the President pro tempore of the Senate and the Speaker of the House of Representatives his written declaration that he is unable to discharge the powers and duties of his office, and until he transmits to them a written declaration to the contrary, such powers and duties shall be discharged by the Vice President as Acting President.

Section 4.
Whenever the Vice President and a majority of either the principal officers of the executive departments or of such other body as Congress may by law provide, transmit to the President pro tempore of the Senate and the Speaker of the House of Representatives their written declaration that the President is unable to discharge the powers and duties of his office, the Vice President shall immediately assume the powers and duties of the office as Acting President.

Thereafter, when the President transmits to the President pro tempore of the Senate and the Speaker of the House of Representatives his written declaration that no inability exists, he shall resume the powers and duties of his office unless the Vice President and a majority of either the principal officers of the executive department or of such other body as Congress may by law provide, transmit within four days to the President pro tempore of the Senate and the Speaker of the House of Representatives their written declaration that the President is unable to discharge the powers and duties of his office. Thereupon Congress shall decide the issue, assembling within forty-eight hours for that purpose if not in session. If the Congress, within twenty-one days after receipt of the latter written declaration, or, if Congress is not in session, within twenty-one days after Congress is required to assemble, determines by two-thirds vote of both Houses that the President is unable to discharge the powers and duties of his office, the Vice President shall continue to discharge the same as Acting President; otherwise, the President shall resume the powers and duties of his office.

AMENDMENT XXVI

Passed by Congress March 23, 1971. Ratified July 1, 1971.

Note: Amendment 14, section 2, of the Constitution was modified by section 1 of the 26th amendment.

Section 1.
The right of citizens of the United States, who are eighteen years of age or older, to vote shall not be denied or abridged by the United States or by any State on account of age.

Section 2.
The Congress shall have power to enforce this article by appropriate legislation.

AMENDMENT XXVII

Originally proposed Sept. 25, 1789. Ratified May 7, 1992.

No law, varying the compensation for the services of the Senators and Representatives, shall take effect, until an election of representatives shall have intervened.

INCIDENT AT SHKIN

Written by Capt Bob Schoultz

I

Their mission was to confirm or deny the presence of Taliban and/or Al Qaeda leadership and prepare for a potential raid on the compound. The Special Forces had been sent in to observe the compound near the village of Shkin after a predator unmanned aerial vehicle had observed suspicious activity at the compound, flying unseen and unheard for a full day overhead. But the Special Forces, like the predator, were unable to tell just what was happening in the compound. Cpt Smith, the team leader of the Special Forces team on the site, had been able to confirm the presence of what appeared to be Al Qaeda and/or Taliban soldiers, but had not yet seen signs of 'leadership.' They had observed armed patrols leaving the compound, moving around the perimeter and then re-entering the compound. They had also observed other activity indicative of a 'military' presence—an armed sentry inside the compound and at least one gathering of armed men in what appeared to be some type of formation. The roof-mounted microwave antennae and mast-mounted antennae in the compound were other indicators that this was not your standard Afghani farmer's compound, housing a large extended family. So they waited and watched.

Intelligence indicated that this old compound was more than merely another military outpost for Taliban or Al Qaeda; it had the potential of offering up some key Taliban or Al Qaeda leaders. Earlier reports had indicated that it had been used as a meeting place for high-level leadership, and being only seven and a half kilometers from the Pakistani border, it was a very convenient staging area for equipment and men, entering and leaving Afghanistan through Pakistan. At one point, intelligence indicated that Osama bin Laden had been scheduled to meet with his doctor at this compound, but the intelligence was determined to be unreliable. It was clear that the compound was a potentially significant Taliban/Al Qaeda outpost and merited close observation.

II

Early on the night of 13 January, after several days on site, Cpt Smith and his team finally observed something unusual, and reported it immediately. Previous Special Forces Teams observing the target had observed regular vehicular traffic coming into the village of Shkin from the border with Pakistan. On this particular night, they observed headlights of a single vehicle leave the specific compound they had been watching, move to the Pakistani border, and flash its lights. In the clear mountain air, the SF team was able to clearly distinguish the headlights of twelve vehicles return down the steep mountain roads from the border to the vicinity of the compound. Several of the vehicles remained at the compound, while others dispersed to other compounds in and around Shkin. Cpt Smith watched carefully for any other unusual activity. It was difficult; at night and from over 2 kilometers away, they couldn't see much, even with their night vision devices. In this part of Afghanistan, by ten o'clock at night, there was scarcely a light to be seen in the whole region, and this night was no different. Soon after the vehicles arrived in Shkin, Cpt Smith and his men were not able to observe any activity—all was dark.

Unbeknownst to Cpt Smith, his report of this unusual activity had received a lot of attention back at Central Command in Florida, and staff officers at some level decided to take the initiative.

III

Cpt Smith's boss was a Navy Seal, Captain Hansen, who was in charge of the Joint Special Operations Task Force (JSOTF), located in Qandahar, responsible for Special Operations support to operations in southern Afghanistan. Capt Hansen was aware of the potential importance of Cpt Smith's most recent report, but was surprised when approximately an hour and a half after getting the report from Cpt Smith, he was contacted by Central Command (CENTCOM). His staff was notified that a B1 bomber was en route with precision guided munitions, with the mission to strike the compound and the

Reprinted by permission of the author.

concentration of Taliban and Al Qaeda vehicles located there. The bombs were scheduled to be on target in two hours. The CENTCOM planners had asked him and his staff how quickly he could put a team on target after the strike to exploit it for intelligence value.

Capt Hansen quickly got his key staff and commanders together and determined that they could potentially put a team on the target a few hours after the strike. It would be difficult and risky, given the very short notice and the distance between where his forces currently were located and the Shkin compound. He would have to redirect forces already planning or conducting other assigned missions, and they would have little time to plan or prepare. But it could be done. However, as he and his staff discussed this surprising development, they sensed that this mission just didn't make sense. He knew from the reports that there were numerous noncombatants living in the compound, and while he also suspected that the vehicles that Cpt Smith had reported could be very significant, they still didn't know who was in those vehicles or what they were doing. Also, he knew that there was no "large" concentration of vehicles at the Shkin compound as the CENTCOM staffers had indicated. They must have misunderstood the report from the field.

Capt Hansen decided that he needed to talk to someone in authority back at CENTCOM to make sure they knew what they were doing. After his communicators got him the secure connection to Tampa, a sergeant at CENTCOM advised him that the only person of authority in the headquarters at the time was Brigadier General (BG) Jones, and he was not available, being in the middle of an important briefing. Capt Hansen directed the sergeant to interrupt the General and tell him that Capt Hansen urgently needed to speak to him.

Within two minutes, Brigadier General Jones was on the phone. Capt Hansen explained to him what was happening, and shared with him his concerns about the mission that was underway. BG Jones was surprised and was completely unaware of the attack that had just been directed. He agreed with Capt Hansen that this strike was inappropriate given the lack of specific information concerning the target, and said he would take action to cancel it. But, he said, the vehicles that were observed may be very significant. He asked Capt Hansen if he could get his forces into the compound and exploit the potential opportunity. It was certainly possible that someone of importance had arrived that night.

Capt Hansen responded that they had a plan on the shelf for an assault, that they would dust it off and make sure it still fit the circumstances. He believed they could assault the target, capture and/or neutralize any resistance with acceptable risk, but he would have to get back to him. BG Jones then had to break off the conversation, explaining that he had to hurry to cancel the airstrike. The strike was cancelled thirty minutes prior to its scheduled time-on-target. Very soon thereafter, CENTCOM sent the JSOTF an execute order to conduct a raid against the compound as soon as feasible.

When he had been asked to put a team on the compound to exploit the planned airstrike, Capt Hansen had hesitated. Cpt Smith's SF team on site, did not have enough men or firepower. He would have had to divert another team, and they would have had to go into the compound unrehearsed and poorly prepared. It was doable, but very risky.

But an assault against the compound was different from putting a team onto a target that had just been bombed. An assault against an enemy position is always very risky. Capt Hansen decided that he would delay until the following night, while Cpt Smith stayed on target and watched. If Cpt Smith observed anything critical, he could speed up the time line if necessary, but his forces needed the extra 24 hours to reposition troops, put the team together, study the terrain and the intelligence, and make sure they went in ready. By waiting 24 hours, he was risking the possibility of missing any leadership that might have arrived that night, but he significantly reduced the risk to his own forces of executing an inadequately prepared and rehearsed plan.

IV

As the strike plan came together, a total of 60 men and 6 helicopters would assault the target. Prior to the assault, Cpt Smith and his team would move in closer to the target to better observe and report on activity in and around the compound, and would watch and report any threats they might see during the raid. By that night, all was ready. Insertion onto the target was scheduled for 2200. The helicopters departed Qandahar and flew together most of the distance to the compound, and then separated about 10 miles out, to allow each to fly at low level his own separate route, to land simultaneously at each helicopter's designated Helicopter Landing Zone (HLZ) around the compound. As the helicopters approached the compound, three of the helicopters missed their designated HLZ's due in part to a navigational equipment error, and due in part to 'brown outs' from dust raised on approaching their HLZ's, which caused the helicopters to drift several hundred yards while trying to find better landing spots. As a consequence, only one of the six SF Teams was able to achieve the surprise they desired.

Immediately upon landing, this one SF Team exited its helicopter and approached the compound to breach an entry. The twelve men burst into the compound, and immediately encountered numerous hostile personnel who had been awakened by the helicopters and the noise. The OIC verbally took control of this group with a small security element while the rest of his team began clearing the buildings inside the compound. On entering their first building, they had to fight hand-to-hand to control and detain two combative males, who were found with two females and a number of children in a room with posters on the wall of Osama bin Laden, and a large quantity of weapons. They found a wide variety of weapons to include mortars, Rocket Propelled Grenades (RPG), a wide variety of ammunition and a 4X4 vehicle. As they continued searching, they discovered a bunker with a large amount of ammunition and anti-tank mines.

Meanwhile one of the other teams had landed so far away that, after exiting the helicopter, they entered what they thought was an outbuilding of the compound complex, and found themselves in a mosque with one person inside praying. The SF soldiers 'flex-cuffed' this individual with the plastic handcuffs they carry, and left him there, while they reoriented themselves and moved on to the compound. On their way to the compound they saw two people hiding in a ditch sprint away into the darkness and escape. They did not fire and continued moving to the compound.

Finally reaching the compound, almost 10 minutes after the first team had arrived, they breached the closed northern gate and entered, moving immediately to buildings nearest their gate. The door to the first building was locked, and they had to breach their way through it. Inside they found a large box covered with a blanket. They moved immediately into the next room where they encountered seven women and six young children sitting against the far wall of the room. The SF soldiers secured the women and children with flex cuffs and returned to the first room and opened the large box. The box contained a wealth of documents, passports, photos as well as numerous AK47's, RPG's, mortar tubes with sights, and a collection of old rifles.

Leaving several men to guard the box and the women and children, the SF soldiers continued clearing buildings. They found the door to the next building locked, and as they prepared to breach the door with a shotgun, one of the soldiers noticed through a crack in the door what appeared to be a woman on her knees on the other side of the door, listening. The soldier stopped the breacher from shooting the locking mechanism, which would likely have killed or seriously wounded the woman. Two men were then able to mechanically breach the door—that is, with crow bars and force—and found the woman with a number of children and three men sitting next to a large safe. One of the men was hostile and combative and required one of the SF sergeants to wrestle him into submission, while his fellow assaulters ensured that the other two males remained passive. The safe was found to contain 198,000 Pakistani rupees, two AK47's, four RPG's, and binoculars.

Another of the SF teams had also landed a good distance from the compound, and struggling with the rough terrain in the darkness, approached the compound as quickly as they could. On their way to the compound, they came upon an individual who immediately fled. After yelling and firing a warning shot the individual stopped, was searched and flex-cuffed. As they neared the compound, they came upon an outbuilding and prepared to enter it. Prior to entering a room

or building, soldiers frequently throw in a "flash-bang"—a small explosive charge which creates enough light and noise ("flash-bang") to temporarily stun and disable anyone in the room. In the first building they entered, a flash-bang managed to set the building on fire and it burned to the ground, but not before four males ran from the building. Two were able to get away by leaping into a ditch and disappearing into the dark, but two were caught and put up a violent struggle before being subdued.

The SF Team then proceeded to and entered the compound and began clearing buildings that the other SF Teams had not yet cleared. In the first building they entered, they found numerous documents along with weapons, women, children, a young man and an older man. The older man became very combative and had to be subdued.

One thing that surprised the SF soldiers during this mission was that women knew to immediately put their hands together in front of them to be flex-cuffed whenever the soldiers approached them. During previous assaults in Afghanistan, women usually became hysterical upon seeing US soldiers and were frequently combative and resistant. They had been told to expect to be raped by the Americans, and to have their children taken away from them. Apparently the word had quickly circulated within Afghanistan that the US soldiers would not rape nor mistreat women nor their children, but would simply flex-cuff them, which they were more than ready to accept.

Once the compound was secured, the FBI agents who had accompanied the soldiers identified the men they wanted to further interrogate, and hung a chem.-lite around their necks, to distinguish them from those to be left behind. The SF team collected all of the munitions and weapons for the Explosive Ordnance Disposal team to destroy prior to extraction, which they did, resulting in a large explosion. The helicopters were then called back in, and at 2315, a little over an hour after the helicopters had landed to insert the assaulters, they took off, taking with them seven males who the FBI had determined warranted further interrogation.

V

Prior to the assault, Cpt Smith and his team had moved their Observation Point (OP) closer to the compound in order to better provide support to the raid. After the assault, Cpt Smith and a small number of his men were directed to stay behind to continue to watch the site for any further developments. The day after the assault, as they watched the locals carefully approach the compound to investigate the events of the night before, they saw one group of what appeared to be local farmers head into the hills in their direction. As they passed within one hundred meters of their position, the locals spotted the OP and approached. The locals were clearly agitated and began gesturing angrily and aggressively, signaling that they wanted the SF soldiers to leave. Cpt Smith then had one of his snipers stand up and point his weapon at the approaching men, which had the intended effect; they calmed down, became more conciliatory and approached the position in a friendly manner. Ever watchful and suspicious, Cpt Smith had one of his sergeants who spoke Arabic try to talk to them. No dialogue or communication was possible, since these Afghanis did not speak Arabic. As the farmers left, Cpt Smith reported the contact to higher headquarters. A short while later, as he and his men were preparing to move their OP, they observed the farmers returning; they had brought with them one of the village elders who spoke Arabic. He had come to bargain, offering to house, feed, and provide water to the men in the OP, if they promised to never bomb his village. Later that night, Cpt Smith and the rest of his men were extracted by helicopter back to the headquarters.

VI

The post operation analysis indicated that the operation against the compound at Shkin was an intelligence coup. Though the operation had not captured key Taliban or Al Qaeda leadership, the prisoners and documents that they did capture proved of great value for later operations.

QUESTIONS FOR DISCUSSION

1. Did Capt Hansen have a moral or legal obligation to contact CENTCOM and inquire about the planned strike?

2. Should he have been held accountable had he not made that effort?

3. Did CENTCOM assume too much risk in choosing to put ground forces against this target instead of bombs? What could have gone wrong? Why did they choose to assume this risk?

4. If the B1 had not been stopped and the compound had been bombed, would this have been a war crime? Why or why not?

5. If it were later determined that Osama bin Laden had been in the vehicles coming from the border, and been at the compound on the night of the scheduled strike, would you still agree with the decision to cancel the strike? What about Capt Hansen's decision to wait 24 hours to conduct the raid?

6. The SF soldiers had many opportunities to shoot to kill. Why were they reluctant to do so? Did they assume too much risk?

7. What were the advantages of conducting this operation the way it was conducted? What were the disadvantages? What were the risks?

INTERDICTION IN AFGHANISTAN

Written by Capt Bob Schoultz

I

It was 2000 hours, March 2002 in the Joint Special Operations Task Force (JSOTF) headquarters in Afghanistan, and LCDR Reynolds had just returned from the chow tent where he had lingered talking with some of the other officers on the JSOTF staff. LCDR Reynolds was a Seal officer in charge of the Seals assigned to the JSOTF conducting Special Operations during Operation Enduring Freedom. Upon returning to the headquarters building to catch up on paper work and review intelligence reports, he was summoned by the JSOTF Operations Officer, LTC Thompson, who wanted to talk to him about a mission they had just received.

LTC Thompson handed LCDR Reynolds an intelligence report and a copy of an email that had just arrived from the Operations Officer of the Land Forces Component Commander (LFCC). The email directed the JSOTF Commander to provide a concept of operations for interdicting a vehicle convoy of Al Qaeda and Taliban terrorists that was expected to be moving down a road about 70 miles to the south the next morning sometime after 0730, apparently trying to escape Afghanistan into Pakistan. It was believed that the convoy might include some key Taliban or Al Qaeda leadership. The LFCC wanted the mission concept in two hours. This meant essentially that the staff wanted to know if the JSOTF thought they could undertake the mission, what support they would need, and whether their plan could be deconflicted with other ongoing missions. LTC Thompson had already contacted MAJ Mark Wyatt, the XO of the Army H47 Helicopter squadron who would be over momentarily to look at the mission with LCDR Reynolds and his men. The mission was to interdict the convoy, and to capture if possible or kill if necessary any suspected members of Al Qaeda or Taliban who they might encounter.

II

LCDR Reynolds knew he had limited time to plan, rehearse, and go over contingencies with his team. Tight time lines had become standard, but they were all fully aware of the increased risk they assumed when they had less time to prepare. A tight time line meant less time to consider and plan for the numerous 'what ifs,' to carefully check the intelligence, and make sure that everyone knew the plan and its various 'branches and sequels.' A recent tragedy could at least in part be attributed to a very abbreviated planning and rehearsal time line. In a high risk,

high stakes operation, a Seal reconnaissance team had been ambushed on insertion by Al Qaeda forces who had been undetected during the pre-mission reconnaissance. The team had been surprised, and two of their friends and teammates had been killed as well as a number of rangers, under the relentless fire of the enemy. The deaths of these teammates were fresh in the minds of his men, and had only steeled their resolve to do whatever it took to find and kill these terrorists. But the enemy was not to be underestimated—LCDR Reynolds and his men knew that their planning must be thorough, and in quick reaction, emergent missions, they always had to weigh the trade-off between the opportunities presented by late-breaking intelligence, and the increased risk of a short planning cycle.

The risks to rapid and short-fused planning, however, had taken an ugly twist two days earlier, when LCDR Reynolds and his team had seen first hand the tragic but unintended consequences that can come from fast-paced operations and decisions made with incomplete or inadequate intelligence. Several days earlier overhead surveillance had seen armed men around a walled compound and corroborating intelligence had indicated that this compound would be used for a meeting of high level Taliban officials. A precision guided missile was launched and struck the main building of the compound during the window when the meeting was scheduled to take place. LCDR Reynolds and his men had been staged to go into the compound minutes after the missile struck to gather any intelligence that remained, capture and treat any wounded, and to determine whether any of the dead or wounded were key Taliban or Al Quada leaders. When they arrived, they found the dead on target had been non-combatants—farmers and their families who were living in the compound. The weapons that were found were personal fire arms that virtually all rural Afghanis possessed and carried for self protection. LCDR Reynolds and his men were shaken by the gruesome results of this miscalculation: elderly people, farmers, women, children, with no apparent connection to the enemy. After determining that there was no exploitable intelligence on the target, he and his men returned to base and reported to his superiors what had happened including his dismay at the mistake. He then refocused his efforts on being ready for his next mission. Part of preparing for the next mission involved dealing with the psychic effects of this one; he contacted the chaplain, told him what had happened, and asked him to talk to the men. Afterward, he knew that having the chaplain meet with them had made a difference, to some of the men more than others, but it felt like the right thing to have done after witnessing, and in a sense participating in the tragic consequences of a mistake in war.

III

After receiving the mission to interdict the convoy from LTC Thompson, LCDR Reynolds knew what to do and started going into his mission planning routine, which had become almost automatic. He was the mission commander, and MAJ Wyatt and the H47s would be under his tactical command. This was just like the seemingly hundreds of exercises he'd conducted, and similar to many of the missions he'd recently conducted during this war. The years of training were paying off. His team was gelling into the type of unit he and every other military officer wants to lead: they only needed to be pointed in the right direction, with a good mission concept and clear commander's intent, and then the plan and preparation just seemed to come together. If everything went as planned and as rehearsed, his role in the execution of the operation would be minimal—communicate with higher headquarters and keep the squad leaders informed about any new developments, and let the squad leaders execute the plan. But of course, nothing ever goes exactly as planned, and it would be his job to make immediate adjustments to whatever unforeseen circumstances they would find, and understand the ripple effect that changes to the plan inevitably caused. That was what he got paid for.

Intelligence indicated that ongoing allied operations were putting significant pressure on Al Qaeda and Taliban forces in Southeast Afghanistan. This increasing pressure was making local Al Qaeda and Taliban movement and operations more and more difficult. Allied forces had received an intelligence tip that some senior leaders, with a group of their armed supporters, would be attempting to escape into Pakistan by vehicle soon after first light the following day. The enemy had already realized that allied aircraft routinely and easily targeted vehicles moving at night; consequently, the terrorists were now seeking to blend in with the normal daylight traffic on the roads. It appeared that Taliban and Al Qaeda were having some success in escaping into Pakistan blending in with the stream of refugees coming out of Afghanistan.

The intelligence indicated that a convoy of three vehicles would be leaving a particular village the next morning and moving toward Pakistan. The vehicles would be SUV's of the Toyota Land Cruiser type and/or compact pick up trucks full of people traveling south on the one road leading to Pakistan. Intelligence sources indicated that normally, the terrorists put their heavily armed men in lead vehicles as an armed reconnaissance element, while the leadership with their personal armed guards would follow some distance behind, maintaining communications with the lead vehicles about any difficulties encountered. Also, and particularly worrisome, were the indicators that the terrorists were probably carrying "Man-portable Air Defense Systems" (MANPADS), specifically, Soviet-era SA-7 shoulder-fired missiles, which are particularly effective against helicopters, especially during daylight when helicopters can easily be seen.

In short order, his men had worked out a plan with MAJ Wyatt and his team. Also the intelligence planners had coordinated with the assigned overhead surveillance; Navy P-3 aircraft would be watching the road and it would be their mission to find and track the targeted vehicles. A very difficult part of the mission was to 'interdict' the convoy in such a way as to achieve complete surprise, while still offering the opportunity for the occupants of the vehicles to surrender without putting his own men at risk. "Capture if possible, kill if necessary" is always tricky, and frequently requires a split second decision, some clear indicator of hostile intent, but also an intuitive sense of threat. But capturing the occupants would be a great coup; he and his men knew that the key to unraveling the terrorist network in Afghanistan was intelligence, and the people in this convoy represented a potential gold mine of intelligence. The Seals would capture them if they could, but if the terrorists resisted with lethal force, as they usually seemed to do, then the Seals were to shoot to kill.

IV

LCDR Reynolds went to see COL Smith, the JSOTF commander to discuss his perspective or any limitations he might have for this mission. With the tragedy of the mission a couple of days previously still on his mind, LCDR Reynolds also wanted to know how certain they were of the intelligence, and whether the rules of engagement had changed. The rules of engagement define the circumstances under which lethal force can be used, and what are the restrictions in the use of that force. COL Smith replied that he understood the intelligence to be quite reliable and the rules of engagement hadn't changed. If the vehicles they encounter demonstrate hostile intent, by displaying or firing weapons, they are legitimate targets. COL Smith believed that the reason higher headquarters wanted the JSOTF to send helos and Seals to do this mission, rather than targeting them from a distance, was because of the desire not to repeat the mistake of two days ago, with which LCDR Reynolds was only too familiar. That said, he reminded LCDR Reynolds that his tactics had to take into account the desire to bring back prisoners if at all possible, while not taking undue risk. In other words, bring back prisoners if you can, but not if it means taking significant risks with the lives of any of your men. COL Smith reiterated to LCDR Reynolds that the rules of engagement gave him all the guidance he needed.

That was what LCDR Reynolds wanted to hear. He felt the rules of engagement as they stood made sense, and gave him and his team the latitude to exercise their professional judgment to complete the mission and stay alive. Rapid assessment of hostile intent in a fast moving tactical environment is a standing requirement, and they had rehearsed and talked through a wide variety of situations many times. He and his men knew the value of prisoners, but they also knew the value of aggressiveness and firepower to staying alive in a gunfight. Their tactics, their survival, and their mission success depended on "Surprise,

Speed, and Violence of Action"—there was no room for timidity. Yet they had recently witnessed the tragic results of "Surprise, Speed, and Violence of Action" exercised without good judgment—in other words, aggressiveness and firepower misapplied.

V

The plan came together quickly—it had to. MAJ Mark Wyatt would be the lead helo pilot for this mission LCDR Reynolds would be in his helo. There would be a total of three helos, referred to as chalk one (with MAJ Wyatt and LCDR Reynolds), and chalks two and three which would carry the rest of the Seals, led by LCDR Reynolds' Assistant Officer in Charge and Platoon Chief respectively. They talked through the contingencies with the pilots and went over the map, and had the intel guys coordinate with the P3's doing the overhead surveillance.

The plan was submitted and quickly approved. The plan was simple and made sense, and at any rate, there was little time to debate it. Their plan had them taking off at 0645 the next morning and flying to a point near the road where they would loiter at a low altitude, visually and audibly sheltered from the road by the mountains, and wait for a cue from the P3 watching the road. When the P3 saw what appeared to be the convoy, it would notify the helos, and vector them to the vehicles on the road. The helos would then move in under the cover of the mountains and surprise the convoy, quickly determine whether to take the vehicles under fire, or if in doubt, land and put the Seals on the ground, and let the Seals make the final determination. The helos would be available to provide cover fire or extraction, as required.

Everyone was very aware of the threat of shoulder fired SA-7's, to which the helos were very vulnerable. An SA-7 missile, in the hands of a reasonably proficient operator, could spell disaster. In daylight, however, helos are also easy prey and vulnerable to small arms fire, and bullets from an AK47 can puncture the skin of their aircraft killing and wounding pilots and passengers. A couple of lucky shots from an AK47 can also bring down a helo and kill everyone on board. As the events in Mogadishu and "Blackhawk Down" had made clear, being in a low-flying helo, near the enemy in daylight is very risky business.

VI

Early the next morning, all went as planned. LCDR Reynolds even got a couple of hours of sleep prior to his meeting at 0530 with his squad leaders and the helo pilots, to go over the plan and review details, one final time prior to launch. The Seals embarked the three H47's, and after all systems checked out and the pilots had established communications with the P3, they took off and headed for the designated loiter point. After about 40 minutes of flight time, they arrived at the loiter point, again checked in with the P3 and began flying in low slow circles, far enough away from the road so as not to be heard, yet close enough to respond quickly when called by the P3.

LCDR Reynolds had been through this drill many times before. Sitting in the helo, with the headset on, partially listening to the relaxed banter of the pilots, he was lost in his own thoughts with the muffled hum and shake of the helo in the background of his awareness. Waiting for the call. Waiting. He mentally walked through the plan for the operation and its various contingencies; how they would make their approach to the convoy, how quickly they would have to determine threat level and response. How far back would the trail vehicle be with the so-called leaders? Would they stumble upon one of the key leaders of the Taliban or Al Qaeda? Did they really have SA-7's?

He pushed from his mind what would happen if the bad guys could get off a shot at the helos with an SA-7 before they could be neutralized. Worrying about it wouldn't do anything. He knew the pilots were very concerned as well; they had discussed it during the planning. But LCDR Reynolds also knew they had a lot going for them on this op—the confidence and skill that comes from extensive training and lots of experience. Surprise, Speed, Violence of Action—their keys to survival, the keys to success.

Approximately 20 minutes after arriving at the loitering point, LCDR Reynolds heard on the head set that the P3 had spotted what appeared to be the target convoy: two pickup trucks traveling together, followed about a mile back by another pick up truck. It would be about 20 minutes before the vehicles reached that section of the road where LCDR Reynolds and the helo pilots had determined that the terrain gave them the greatest advantage for surprise, and the bad guys the least opportunities for escape, on vehicle or on foot. After discussing it briefly with MAJ Wyatt, LCDR Reynolds advised the Seal Leading Petty Officer (LPO) in his helo what he had just heard, and the LPO alerted the rest of the Seals. The Seals then seemed to come alive. Up to that point, they had been sitting in the back with their eyes closed, some probably dozing lightly, some probably rehearsing the mission in their heads, some probably thinking of things completely unrelated to this operation. But now all the men were alert and focused, checking their gear one more time, adjusting their position to be better prepared to exit the helo in a hurry.

MAJ Wyatt continued to get information from the P3. The convoy was continuing down the road toward the interdiction point. After about 10 minutes, the P3 crew advised MAJ Wyatt that it was time to leave the loiter position and begin moving toward the road. LCDR Reynolds advised his LPO and the LPO passed it on to the men in the helo.

As the helos approached the interdiction point, they stayed very low to the ground, flying at about 50 feet, to minimize the chances that the "wop, wop, wop" of their approach would get over the mountains and alert the convoy. At about 2 minutes out, the P3 passed on some disturbing news. "We've lost the trail vehicle. We haven't seen it for several minutes—last we saw it was about 3 miles back. It might be masked by the mountains between us and them. But two vehicles are on final into your target zone and will be there in a couple of minutes."

"Damn!" LCDR Reynolds thought. Quick decision time. The plan had been for him and MAJ Wyatt to break off from chalks two and three in the last twenty seconds, and to go to the trail vehicle, to permit a simultaneous hit on the lead and trail vehicles. He was going to the trail vehicle, because that was where the real valuable targets would be—the leaders. LCDR Reynolds quickly considered the possibility of his helo flying thru the mountains searching for the trail vehicle while chalks one and two were taking care of the lead vehicles. There was no telling where that vehicle could be or what it could be doing. Even though the primary target was the leadership in the trail vehicle, with this new uncertainty, LCDR Reynolds did not want to take off on a potential wild goose chase, splitting his force, now that the plan seemed to be coming unraveled at the last minute.

He told MAJ Wyatt he wanted to keep all three helos together until they had a better idea what they were up against. Or at least until the P3 found the third vehicle. MAJ Wyatt concurred and told the chalks one and two that the plan had changed and that they would stay together and all hit the lead vehicles. They then started their climb up and over the final hill that lay between them and the road, and presumably the two lead vehicles. LCDR Reynolds ensured that the word was passed to the Seals in chalks two and three. Everyone in the helos was on full alert, the pilots and crew calmly passing information back and forth, the Seals on their feet, looking out the windows, weapons at the ready, on safe.

VII

As they popped over the summit of the hill, they saw about five hundred feet below them and to the left, two pick up trucks approaching from the north. LCDR Reynolds suddenly experienced that familiar jolt of adrenaline, a combination of stress, excitement, responsibility and complete focus. The helos came over the crest of the hill and headed down low and fast, directly toward the vehicles, approaching at full speed, circling from left to right, counter-clockwise. LCDR Reynolds stared intently at the occupants in the back of the pick up truck, looking for any sign of hostile intent. First the front vehicle, and then the rear vehicle stopped when they saw and heard the helos, and he saw men get out and begin running. Then LCDR Reynolds thought he saw weapons and muzzle flashes. LCDR Reynolds was looking over the shoulder of the left door gunner, who also saw the weapons and muzzle flashes, and immediately opened up on the lead vehicle with his mini-gun, shifting to the second vehicle as soon as he could get a good shot at it. At about that time, the second helo picked up the lead vehicle and started cutting it to pieces.

LCDR Reynolds saw more muzzle flashes and then saw men fall. No sign of anyone setting up to fire an SA-7. The helos passed the vehicles flying fast and low and putting out a huge volume of fire. The two pick up trucks were being cut to pieces, and men who had not been able to get out of the vehicles in time were being chewed up as well. Those who had left the trucks were scrambling in chaos and disorder, some firing at the helos, several of them falling victim to the withering fire coming from the door gunners. LCDR Reynolds saw that this part was going well. Now, where was the trail vehicle with the leadership?

As his helo was turning to circle the vehicles and make an approach from the other side, LCDR Reynolds felt that chalks two and three could handle this. He said to MAJ Wyatt on the headset, "Mark, I think they've got this under control. Let's go find the trail vehicle. What do you think?" "Roger," he responded. " I'll advise chalk two to take control here," at which point he pulled up out of the pattern and told the pilot of chalk two that he and LCDR Reynolds were detaching to go look for the other vehicle. The P3 had just called to tell them that they still had no sign of the third vehicle. MAJ Wyatt told the P3 what he was doing, and then he turned and headed up the road down which they had seen the two vehicles coming.

VIII

LCDR Reynolds called his LPO up to him, took off his head set, and yelled over noise of the helo to tell him what they were doing. The LPO nodded and then went to the back of the helo to tell the other Seals who, still very tense and focused, were looking toward him with some anticipation. They knew that something was up. LCDR Reynolds then moved to the door gunner on the right side of the aircraft, since the helo was flying with the road on the right side. The longer it took to find the vehicle, the greater the risk. They had to assume that the trail vehicle had heard the helos and the gunfire, and perhaps even had radio communication from the lead vehicles. That gave the bad guys plenty of time to set up on the helo—they would certainly be expecting them. These were the leaders, and they would have the most devoted soldiers with them as bodyguards, and probably the best weapons, possibly to include SA-7's. Helos are big, easy targets in daylight, especially if you know that they are coming. The right door gunner had not expended any ammunition on the assault on the other two vehicles—he was keyed up, ready, and had a full load of ammo.

As the H47 flew down the narrow valley that hugged the road, there was an intense and anxious silence on the headsets. The pilots, crew and LCDR Reynolds knew that this was where they were most vulnerable. Though they might not achieve complete surprise, they hoped to overwhelm the bad guys by hitting them suddenly and with overwhelming firepower. But they had to be lucky and good.

As the H47 turned a corner in the valley, they looked up a narrow canyon. LCDR Reynolds saw the pick up truck just as he heard MAJ Wyatt calmly say, "There they are." What looked like a truck full of people was stopped on the side of the road about 200 yards ahead to the right. The door gunner had a clear shot, and he quickly swung his mini-gun and took aim.

IX

LCDR Reynolds suddenly sensed something wasn't right. Just as the truck came into view, just as the door gunner swung his weapon in the direction of the truck, just as MAJ Wyatt said, "There they are," LCDR Reynolds in an instant realized that no one was running from the vehicle, and he thought he saw someone in the truck (a woman?) hold something up high as if to display it to the helo. He grabbed the door gunner and yelled "NO!" and held his fist in front of the door gunner's face in the signal for "Stop what you're doing!" The door gunner was confused, but he followed the order and didn't shoot. The helo continued toward the truck, low and fast as LCDR Reynolds looked hard at the truck, looking for signs of hostile intent. In the two long seconds it took to get to and pass the truck, they noticed that this was different from the other vehicles. No one left the truck. No one ran for cover. It was hard to tell whether these people were armed or not, given the speed and approach angle of the helo. The helo sped past the truck so close that the people in the bed of the truck were ducking from the rotor wash, and LCDR Reynolds saw that he had been right—it had been a woman he'd seen, and what she was holding up appeared to be a baby. He didn't see any weapons yet or anyone displaying hostile intent. That didn't mean they weren't bad guys, and that they weren't a threat. LCDR Reynolds told MAJ Wyatt to circle around and land in front of the vehicle, far enough away to be safe, but close enough for the Seals to quickly envelope the vehicles, clarify the situation, and take appropriate action.

After speeding by the vehicle, MAJ Wyatt exhaled. When he didn't hear the door gunner firing, he thought the weapon had jammed and that they were 'done for.' He flew the H47 at full throttle farther down the road, banked around a bend in the road, and then ascended to fly over a hill to come back to a position several hundred yards in front of the vehicle. He was ever mindful of the possibility that an SA-7 was being prepared for the first clear shot. LCDR Reynolds dashed back to his LPO and told him that the Seals would debark and move in to observe the vehicle—it wasn't clear if these were hostiles. He then moved back to the front of the helo so that he could get oriented prior to landing.

The helo flared and landed fast. The Seals quickly debarked out the rear ramp and moved to outside the rotor-wash to set up a hasty perimeter in the nearest cover. The H47 lifted off the ground, turned 180 degrees away from the direction of the vehicle, and took off. The Seals patrolled to the vicinity of the pick up truck and observed the passengers not moving, sensing their danger. LCDR Reynolds was able to signal to the passengers to move away from the pick up truck. He then had his team search the pick up truck and its passengers, and determined that they were not Taliban nor Al Qaeda leadership, nor was there any evidence that they had any connection to them. Either the intel had been wrong about the three-vehicle convoy, or the situation had changed since the source had reported it. It didn't matter. These people did not fit the profile of Taliban or Al Qaeda and happened to be in the wrong place at the wrong time.

LCDR Reynolds realized that he had narrowly avoided making a tragic mistake. He was still worried about a possible trail vehicle, and called MAJ Wyatt to ask him if he had any other information. MAJ Wyatt had been in touch with the P3, and had gone to altitude himself to see if he could see any other vehicles, and there was nothing. LCDR Reynolds then got on the radio with the Seals who were on the ground at the site of the two lead vehicles. They had already debarked the helos, taken control of the site with no resistance, and they were inspecting the dead and wounded. All were males and had been carrying arms. Eight were dead, the three wounded were being treated, and they had taken two unscathed prisoners, who had survived the initial assault, and had stood with their hands raised when the Seals approached. This was all good news.

LCDR Reynolds then had his LPO direct the civilians to sit down and to remain where they were. They were still sitting on the ground away from their pick up truck when the Seals were picked up by the helo and flown to join their teammates at the site of the two lead vehicles.

Part C:

COL Smith, the JSOTF Commander had heard that his helo pilots believed that LCDR Tom Reynolds had taken undue risk during an operation from which they had just returned, and that this was causing some tension between the Seals and Army helo crews. COL Smith had heard what had happened and was familiar with the events of the operation, but knew he needed to get the story directly from his two commanders. He called MAJ Wyatt and LCDR Reynolds into his office to get the issues out on the table.

LCDR Reynolds and MAJ Wyatt walked into his office, and after COL Smith indicated that he understood that there was some disagreement about how the operation had been conducted, MAJ Wyatt, clearly emotional, addressed the issue right up front:

"Sir, we could have all been killed, and lost the bird. We were a sitting duck. We're real lucky Tom was right, because if he'd been wrong, we would have a lot of dead Americans and this war would look a lot different right now." MAJ Mark Wyatt stepped back and exhaled slowly.

COL Smith looked at LCDR Reynolds and indicated it was his turn to speak.

"Sir, he's right—we could have all been killed—if I'd been wrong. But I wasn't. I was in charge. And I was right. I made the call based on what I saw, and what I sensed, and I stand by it. It was clearly the right thing to do. We knew we were at risk, but we still have to do the right thing."

Mark Wyatt jumped on him. "Right Tom, but all the indicators were there that these were bad guys, and you didn't KNOW, and my guys and yours were sitting ducks for several seconds, and that put not only all of us, but potentially the whole focus of everything we're doing here at risk. Can you imagine what this task force would be doing right now if those had been bad guys and we had taken an SA-7 right down the throat? I don't want to kill innocent people either, but if you had been wrong, nobody, I mean NOBODY, would forgive you. And we'd all be dead."

"Mark—it just didn't feel right—and, we saw no hostile intent."

"We didn't have time to see hostile intent, Tom! When we took off after that third vehicle, my understanding was that we were going hunting. We knew we had flushed the bad guys, and at that point, we were in a gunfight. When we came around that bend in the road, it was either them, or us. When you stopped my gunner, and I didn't hear the guns, I figured it was us. I expected a flash and woosh and then lights out."

"You two calm down and come back and see me when you get your stuff squared away," interrupted COL Smith. He knew that he was the one who had to take responsibility for risk, and if there was something unclear about risk, he needed to resolve it. "I'm going to have to think about this, and talk to the lawyers. Now get out of here and get some rest. We've got a bunch of other things hopping and we'll need you to be focused."

MAJ Wyatt and LCDR Reynolds left the Colonel's office and agreed to get together in a couple of hours after they had taken care of their men and their gear, and sorted out the other details from their mission. MAJ Wyatt was clearly still upset as he walked away to rejoin the other pilots preparing their reports.

As LCDR Tom Reynolds walked back to where his men were working, he thought about what his friend Mark Wyatt had said. He had gambled and won, but he had bet the whole farm—not just his farm, but the lives of everyone else in the helo, as well as the future capability of the Special Operations Task Force.

QUESTIONS FOR DISCUSSION

1. List all the points in the case when a decision was made whether to fire, or not fire.

2. What was each of these decisions based on?

3. How does an officer weigh the risk to his/her people versus the risk to non-combatants?

4. How important do you believe LCDR Reynolds' experience at the compound several days prior to this mission was in his decision not to fire? Do you think he would have made the same decision had he not had that experience?

5. Did the attack on the compound violate the Laws of Armed Conflict principles of discrimination and proportionality? Would an attack on the third vehicle, had it occurred, been a violation of these same principles?

6. Did LCDR Reynolds' decision show exceptionally good judgment, was it an obvious call, or was he just lucky?

7. Should LCDR Reynolds be praised for his decision? Are there any dangers to giving too much positive recognition to military officers who make that decision?

8. What were LCDR Reynolds' moral obligations here, to his country, to his troops, to the non-combatants?

9. If LCDR Reynolds had not stopped the door gunner, and had they killed or wounded the non-combatants in the third vehicle, would that have justified a court martial? Are there any legal vs moral issues here?

ACTING ON CONSCIENCE: CAPTAIN LAWRENCE ROCKWOOD IN HAITI

Written by Dr. Stephen Wrage

Port-au-Prince, Haiti
Barracks compound, Tenth Mountain Division
30 September 1994
1920 hours

Captain Lawrence Rockwood, counterintelligence officer[1] with the U.S. Army's Tenth Mountain Division, crouched by his pallet on the concrete barracks floor and thought back through what had happened over the past seven days.

Six days ago, on his second day in-country, a report from the Belair jail in Port-au-Prince described a mutilated Haitian torture victim spirited out at night. A report two days later traced a beheaded body found in a swamp outside the city back to the Omega jail. All the prison reports featured emaciated and abused prisoners, not criminals in most cases—simply enemies of the regime that the American forces were there to replace.[2]

A report he had received two days before on the 28th said that American forces had entered a prison in the southwestern town of Les Cayes. They found "over 30 men were crammed into a cell no larger than 15 feet square. They were so malnourished that—as with concentration camp victims of World War II—their food intake had to be increased gradually to avoid harming them. When the American soldiers removed one invalid from the prison, they discovered that he had lain for so long in one position that some of his skin had fallen off."[3]

"At least we could get food into those places," Rockwood thought. He had seen the pallet-loads of MRE's—Meals Ready to Eat—unloaded from American ships onto the docks in Port-au-Prince. He had even told one prison official he could probably get two per day delivered for each of his prisoners. The official was against it: too great a security risk, he said. "What's the risk?" Rockwood had asked him. "It's the starving prisoners who will riot, isn't it?" No, he was told. The starving ones just lie there. The security risk would come from outside the prison: from all the people who would break in to get at that food.[4]

An Unusual Soldier

Rockwood had arrived in Haiti seven days earlier on September 23, four days after the first American troops were deployed to the island. He had prepared for this mission with eager anticipation. Rescuing the helpless and opposing the tyrannous is precisely what a military is for, he thought.

Rockwood was the son, grandson, and great-grandson of military men, but he didn't fit any traditional mold. He had grown up on military bases—both his parents served in the Air Force so by the time he went to high school he had lived abroad in Turkey, France, and Germany. The event he remembered best from his childhood was when the family was stationed in Germany.

Years before, his father had been among the forces that liberated the Nazi camps. He wanted his son to know what he had seen and learned, so when Rockwood was eight years old he and his father went together to the concentration camp at Dachau. "My father told me that these camps are not the creation of a few evil, brutal men. They're really the creation of cynicism and blind obedience to authority."[5]

Rockwood considered breaking the pattern of three generations by joining the priesthood instead of the military, but after a year in a Catholic seminary he followed suit and enlisted in the Army. He was 19 then, and along his unusual track to being commissioned as an officer he would earn a bachelor's in psychology, a master's in history and become a licensed practical nurse. He would also convert to Tibetan Buddhism. Before his deployment to Haiti he had been treated for depression and at the time of the deployment he was taking the anti-depression drug, Prozac.

In the Army he rose fast and received outstanding evaluations.[6] He chose his models carefully and worked hard to mold himself in their pattern. In his cubicle back at Fort Drum he kept pictures of three men he admired: General George Picard, a counterintelligence officer in the French army during the Dreyfus Affair who went to prison to protest Dreyfus' innocence,[7] Colonel Count von Stauffenberg of the German army who gave his life in an attempt to assassinate Hitler, and Chief Warrant Officer Hugh C. Thompson, the helicopter pilot who saw the My Lai massacre in progress, lowered his helicopter into the middle of it, and ordered his door gunner to train his machine gun on American troops who were killing unarmed civilians.

Rockwood's Concern

Well before he left for Haiti, Rockwood was worried about human rights abuses there, and he focused on Haiti's prisons as the likeliest sites of torture, murder, and abuse. On the 10th of August he requested a special classified report from the C.I.A. about Haitian prisons, and later he would point out that the Civil-Military Operations Handbook for the 10th

Reprinted by permission of the author. Stephen Wrage is a professor of political sicence at the US Naval Academy.

This case was produced by the Center for the Study of Professional Military Ethics, US Naval Academy. Support was provided by Newport News Shipbuilding.

Mountain Division includes a checklist enumerating the information the division staff should obtain about each site where prisoners were confined, including "name, address, grid coordinate, telephone number, type of facility, maximum capacity, present capacity, number of guards, capacity of kitchens, name of warden, overall condition of facility and inmates."[8]

Rockwood was confirmed in his commitment to human rights in Haiti when he heard President Clinton say in his September 15 address to the nation that a primary objective of Operation Uphold Democracy was "to stop the brutal atrocities."[9] He was proud to be part of the team when his unit began to deploy to Haiti on the 19th. He arrived in Haiti four days later.

The Background

For over a year Captain Rockwood had watched the situation in Haiti unfold.[10] As an Army counterintelligence officer stationed at the headquarters of the 10th Mountain Division in Fort Drum, New York, he had monitored the long play of threats and defiance pass between the Clinton Administration and the Cedras regime.

Three years earlier, in September 1991, General Raoul Cedras had overthrown the only democratically elected government in Haiti's history when he drove Jean-Bertrand Aristide into exile only seven months into his term.

Two years after that, in October 1993, a noisy crowd encouraged by Cedras had blocked the docks in Port-au-Prince when the U.S.S. Harlan County with U.S. and Canadian troops, engineers and trainers aboard, had tried to land. Rather than face the prospect of even minor violence, the Clinton administration pulled back the Harlan County. They may have been unwilling to open a new front in the peacekeeping struggles, since the week before that, 18 American soldiers had been killed by Mohammed Aideed's gunmen in Mogadishu. American enthusiasm for nation building was at a low point and the Harlan County steamed back to the United States.

By spring of 1994, however, the Clinton Administration was facing strong pressure to act. The Congressional Black Caucus had publicized torture and murder in Haiti; Randall Robinson of TransAfrica had begun a hunger strike in sympathy with the victims of the Cedras regime; Clinton's chief advisor on Haiti had resigned and been replaced with a former head of the Black Caucus; midterm elections were six months away and desperate Haitian refugees were appearing on the beaches of Florida.

In July 1994 President Clinton sent the 24th Marine Expeditionary Unit to float in the waters off Haiti and threaten imminent force, and in early September Clinton sent forces aboard the carriers "Eisenhower" and "America" to join them, but Cedras remained adamantly in power. On September 15 Clinton at last said "there is no point in going any further with the present policy"[11] and airborne Special Operations units boarded their planes at Fort Bragg. Rockwood monitored the cable traffic and CNN, expecting see what the military calls a "non-permissive entry."

The paratroopers were already in the air when an emergency mission led by former President Jimmy Carter, Senator Sam Nunn and General Colin Powell induced Cedras and his top circle to leave Haiti. American troops led a multi-national force into the country unopposed, but they entered a strange setting. Aristide was not scheduled to return to Haiti for another month. Until then governance was to be shared by the American-led, U.N.-sponsored forces and the remains of the Cedras regime which had proven itself corrupt, brutal, and frequently murderous. The prisons, for example, remained under local control.

Force Protection

"As I assumed my duties in Haiti on September 23 I was informed that 'force protection' was to be the focus of our efforts," Rockwood later reported.[12] This troubled Rockwood and others but seemed entirely appropriate to many members of the mission. Assuring "force protection," avoiding "combatant status" and resisting "mission creep" were the lessons learned from the previous October's disaster in Somalia. Joint Task Force Commander Lieutenant General David C. Meade and his staff officers were determined that American troops in Haiti would not cross "the Mogadishu Line."[13]

When troops landed on September 19, their rules of engagement had required them to stand by or look the other way as thugs from the Cedras regime beat Aristide supporters who had gathered at the port to hail the Americans' arrival. Americans were to use force only when they were themselves threatened with violence; Haitian-on-Haitian violence was not to be resisted. American troops were to stay for the most part behind barbed wire and sand bag emplacements and were forbidden to leave the barracks compounds unaccompanied. Most troops could move about only in convoys of at least two vehicles with at least two persons in each vehicle.

Even though Meade's multinational force had arrived in overwhelming strength—20,000 troops plus heavy equipment"[14] there were many challenges to its authority. Besides the thugs on the docks who beat the pro-democracy demonstrators, there were a number of tense confrontations with unruly crowds. Violent incidents, however, were few. One American soldier was shot by a Haitian he had arrested, and on September 24, when a patrol of Marines were fired on in Cap-Haitien, they returned fire and ten Haitians were killed.

Rockwood's Odyssey

Rockwood was convinced that Haitians, not Americans, were in the greatest danger. "The main content of the reports that reached me centered on human rights violations against Haitian slum residents rather than any threats directed against our forces," he later said.[15] As soon as he arrived, Rockwood embarked on what he called "my week long odyssey . . . to awake interest of the commander and staff of the Multinational Forces in human rights violations."[16]

On the evening that he arrived in-country, September 24, Rockwood called on Lieutenant Colonel Karl Warner, chief legal officer of the 10th Mountain Division and the man responsible for monitoring human rights violations. Since Colonel Warner was not in, Rockwood left a message requesting authorization to look into the National Penitentiary in Port-au-Prince, which he believed to be the site of atrocities.

The next morning Rockwood met with the command's chaplain to speak of the deteriorating human rights situation in Port-au-Prince slums and the particular problem of the prisons. Rockwood reports that the chaplain said he did not want to get involved in a "political" problem.[17] Rockwood remonstrated with him and later made a formal complaint regarding the chaplain's attitude in a letter to the head of the chaplaincy corps.

That same day, September 25, Rockwood went to the staff Judge Advocate's office and asked for the Laws of War manual, the 1977 Protocol to the Geneva Convention or the report on the U.N. High Commission for Human Rights Conference held in Vienna in 1993. He was determined to prove that the Joint Task Force had an obligation under international law to protect human rights in Haiti. He was disappointed to find the only available reading material was an Army field manual compiled in 1954.

Rockwood's sense of urgency was heightened that day as he received the report from Belair jail mentioned above. Late in the day he took that report to his commanding officer, Lieutenant Colonel Frank Bragg, who had been "something of a mentor to Rockwood. Bragg was sympathetic, but said prison inspections weren't a realistic goal. He told Rockwood to focus on protecting U.S. forces, not Haitian civilians."[18]

Rockwood returned the next evening, September 26, to the Judge Advocate's office to protest the lack of action on human rights violations. Rockwood's sense of desperation was growing as he was convinced that the Cedras regime was using its last few days in control of the prisons to eliminate its enemies—political opponents who had been victims and witnesses to crimes of torture and murder.

On September 27 Rockwood called at the Civil-Military Operations Center hoping to spur a survey of the penitentiaries. He was told that the operations center was not collecting current information on the prisons because the Joint Task Force had no jurisdiction there. He offered the reports he had received on the Belair and Omega jails.

That evening he attempted to organize an intelligence team to visit several prisons but was told he would need a military police escort. The military police refused him an escort, saying their orders were to monitor Haitian police stations and police patrols but not prisons.

Rockwood argued to anyone who would listen that a primary principle of intelligence work is to protect your sources, and warned that the people he talked to during the day were disappearing apparently being arrested or killed overnight. He needed to go to the prisons to see if they were there. He was told to be patient. It would be some time before troops could be spared for such missions. On the morning of the 29th a liaison officer from Special Operations Forces called on Rockwood to tell him that Rockwood's unit was to take no destabilizing action, and in particular that they were not to inspect a prison without full military support.

Convinced that innocent people were dying and feeling responsible for their fate, Rockwood grew desperate. Late on September 29 he went to the Inspector General and lodged a complaint alleging that the Joint Task Force command was failing to protect the human rights of people in the territory it occupied and controlled. He named eight officers in his chain of command and charged that they had subverted President Clinton's primary mission intent concerning human rights as announced in the September 15 address to the nation. Under "Action Requested" he wrote, "Inform the commanding general as soon as possible of facts that may lend the appearance that the Joint Task Force is indifferent to probably ongoing human rights violations in the [Port-au-Prince] penitentiary."[19] The Inspector General discouraged Rockwood from approaching the command's Chief of Staff on this matter, but he also told Rockwood that his complaint would not be brought to the attention of General Meade for at least a week.

Rockwood did not go to the Chief of Staff. Instead that evening he again confronted his commanding officer, Lt. Col. Frank B. Bragg, and detailed his concerns. He reportedly compared General Meade to General Yamashita, the commander of Japanese forces in the Philippines in 1945.[20] Yamashita was sentenced to death by a war crimes tribunal for his failure to protect American prisoners, even though he neither ordered nor knew of their execution by his soldiers. General Meade, Rockwood argued, had direct and specific knowledge of human rights abuses in the Haitian penitentiaries, and was doing nothing to stop them. Lt. Col. Bragg had no sympathy with these arguments.

The Decision

Now, several hours after that confrontation, late in that long day of the 30th of September, seven packed days after his arrival in-country, Rockwood got up off the floor in the barracks in Port-au-Prince. He knew what he would do next.

ENDNOTES

1. As a counterintelligence officer, Rockwood's duties were to read intelligence reports and debrief intelligence operatives, both American and Haitian, to discover potential threats to the security of U.S. forces in Haiti. In this role he had unusual access to information, freedom of movement, contact with Haitians and opportunity to exercise initiative.

2. A Central Intelligence Agency report that Rockwood had requested before he set out for Haiti said "85% of the 300 to 500 people incarcerated [in the National Penitentiary in Port-au-Prince] have not been charged" with a crime. The report found they were political prisoners of the Cedras regime, supporters of the democratically elected Aristide government that the intervention was intended to restore to power. See Meg Laughlin, 'The Rockwood Files," Miami Herald, October 1, 1995, Tropic section, page 6.

3. The prison had been visited by special forces operating independently in the countryside under the command of Lieutenant Colonel Michael Jones. In an interview with Bob Shacochis, author of The Immaculate Invasion (New York: Viking, 1999) Jones says, "We found some photographs, pretty damning photographs. People being pulled apart with chains, people being beaten." (Shacochis, p. 150.) Jones later recalled "a pile of live bodies crammed into a cell in which there was neither room to stand nor room to lie down. When soldiers, who apparently did not realize initially that the men were still alive, began pulling one of the men off the pile, his skin simply ripped off his back, exposing his spinal cord to view." Quoted in transcript of U.S. v. Rockwood, no. 261–2–6597 at 1604–5. See also Ian Katz, "Depressed or Just Decent," The Guardian, (London) May 30, 1995, at T4 and Peter Slevin, "36 Inmates, One Cell: Haitian Jails in Squalor," Miami Herald, October 10, 1994, at 1A. The horrible conditions in Les Cayes were not unique. General James T. Hill, deputy commander of the 25th Infantry Division deployed to Haiti in 1996, told reporter Anna Husarska in an unpublished interview, "everybody found it in every one of the jails. There is no doubt about it. I've been to almost every one of the jails." Interview with Husarska dated March 2, 1995. See Robert O. Weiner and Fionnuala Ni Aolian, "Beyond the Laws of War: Peacekeeping in Search of a Legal Framework," Columbia Human Rights Law Review, Winter, 1996 at note #21.

4. See testimony of Paul J. Browne, Vice President, The Investigative Group, in United States House of Representatives, 104th Congress, First Session, Human Rights Violations at the Port-au-Prince Penitentiary, Hearings before the Subcommittee on the Western Hemisphere, Committee on International Relations, May 3, 1995.

5. Quoted in Associated Press, "Court-martial Looms for Officer Who Probed Haiti Rights Abuses," Asheville Citizen-Times, Asheville, NC, at 3A.

6. Officer fitness evaluations of Captain Rockwood between 1987 and 1993 characterized his performance as "superb" or "excellent" and recommended he be promoted "ahead of his contemporaries." Transcript at Defendant's Exhibit U, United States v. Rockwood, no. 261–29–6597 (M.J. 1995).

7. "In 1894 Captain Alfred Dreyfus (1859–1935), a French officer, was convicted of treason by court martial, sentenced to life imprisonment, and sent to Devil's Island. The case had arisen with the discovery in the German embassy of a handwritten list of secret French documents. The French army was at the time permeated with anti-Semitism, and suspicion fell on Dreyfus, an Alsatian Jew. . . . In 1898 it was learned that much of the evidence against Dreyfus had been forged by army intelligence officers." The Concise Columbia Encyclopedia, (New York: Columbia University Press, 1983) page 242.

8. Civil Military Operations Handbook of the 10th Mountain Division, Entry #9, "Law Enforcement Agency Checklist." See also the Civil Affairs Operations manual of the U.S. Army (FM 41–10) at Chapter IX (Public Safety) under heading "c."

9. President Clinton's words: "Our reasons are clear: to stop the horrible atrocities; to affirm our determination that we keep our commitments and we expect others to keep their commitments to us; to avert the flow of thousands more refugees and to secure our borders; to preserve the stability of democracy in our hemisphere." Foreign Policy Bulletin, November/December 1994, page 18.

10. Haiti is a mountainous country of about 11,000 square miles and 9,000,000 people, almost all of African descent. It trails every country in the western hemisphere in such measures of development as literacy, income per capita, doctors per thousand people and miles of roads. 85% of the population is illiterate; 60% are unemployed or underemployed. Less than 40% of the urban population and less than 5% of the rural population have access to piped water. Infant mortality is over 110 per thousand (compared to 40 per thousand in the United States). Brian Weinstein, Haiti: The Failure of Politics, New York: Praeger, 1992, pp. 4–5.

Before 1790, Haiti was France's richest colony, accounting for almost half of France's foreign trade and producing 50% of the world's sugar and 40% of the world's coffee. A series of bloody revolutions in the next twenty years and a brutal but inefficient feudal system throughout the 19th century entrenched Haiti in misery. The country was occupied and governed by U.S. troops from 1915–1934. Since then a succession of dictatorships protected the interests of a wealthy, Europeanized elite at the expense of the mass of the population.

11. See Foreign Policy Bulletin, November/December 1994, page 18.

12. Interview with the author, August 18, 1999. Rockwood was not alone in that assessment. See also the testimony of Lieutenant Colonel Frank Bragg, Assistant Chief of Staff for intelligence, 10th Mountain Division and Director of Intelligence for the Multilateral Force in Haiti: "Question: Would it be fair to say that actually your whole priority was force protection at that time? Answer: It is fair to say that there was no doubt, that was my number one priority and I had every intelligence asset I could muster focused primarily on that one thing." Transcript of U.S. v. Rockwood, no. 261–29–6597 at 1372.

13. On "mission creep," see Adam B. Siegel, *The Intervasion of Haiti*, Professional Paper 539, August 1996, Center for Naval Analyses, page 27.

14. Of those 20,000 troops, about half were in logistical, communications, intelligence, or other support roles. The troops of the Joint Task Froce were primarily concentrated in Port-au-Prince and housed in a converted industrial park on the edge of the city. Small units of special forces operated independently in the countryside.

15. Interview with the author, August 18, 1999.

16. Interview with the author, August 18, 1999.

17. "He said he didn't want to get involved in a political issue. He said he was concerned about morale. . . . It was the most categorical response that I got from any officer." Rockwood to Pinsky in a telephone interview. See Mark I. Pinsky, "Changing Role of Armed Forces Complicates Military Clergy's Task," *The Orlando Sentinel*, 1 December 1996 at G-1.

18. Interview with the author, August 18, 1999. Quotation is from Meg Laughlin, "The Rockwood Files," *Miami Herald*, October 1, 1995, Tropic section, page 8.

19. This series of events is described in Rockwood's testimony before Congressman Dan Burton's Subcommittee on the Western Hemisphere of the House Committee on International Relations. See United States House of Representatives, 104th Congress, First Session, *Human Rights Violations at the Port-au-Prince Penitentiary*, Hearings before the Subcommittee on the Western Hemisphere, Committee on International Relations, May 3, 1995.

20. Interview with the author, August 18, 1999.

*Written by Dr. Martin L. Cook
and Major Phillip A. Hamann*

1. The Facts

On Monday, 25 February (the second day of the ground war), American intelligence agencies passed reports from the Kuwaiti Resistance inside Kuwait City to the military command center in Riyadh that the Iraqi occupation forces were preparing to leave the city. Kuwaiti Resistance reported that members of the Iraqi secret police, the Mukhabarat, were attempting to destroy evidence of war crimes (killing all tortured Kuwaitis for example) and pillaging as much property as possible.[1] The Resistance also boasted of mounting a small offensive against the panic-stricken Iraqis. This offensive did not really amount to much—the resistance movement was known to exaggerate at times. But the resistance movement never suggested that the convoy that was preparing to leave Kuwait City contained kidnapped Kuwaiti citizens. (Most of the kidnapped Kuwaiti citizens had already been sent to Basra and other locations weeks prior to this incident.)[2] Air Force intelligence therefore surmised with high confidence that the convoy consisted exclusively of panicked stragglers from the decimated front line divisions (the Iraqi III Corps) and the Iraqi secret police.

In addition, at the start of the ground campaign, American intelligence agencies intercepted an uncoded telephone message from a general officer of the Iraqi Republican Guard.[3] The officer appeared to issue a general order of retreat to Republican Guard units in Kuwait and to order the setting up of a screening or blocking maneuver to allow the Republican Guard to get out of the Kuwait Theater of Operation and into Basra. Hence, American intelligence was aware of preparation for a sudden and massive exodus of Iraqis from Kuwait.

The Mukhabarat secret police is a paramilitary organization with little access to heavy military armor. Its members frequently do not wear uniforms. Therefore, the fact that some individuals on the road were not uniformed was consistent with their identity as Iraqi secret police.

The evening of the same day, a JSTARS aircraft (a recently modified Boeing 707, with discriminating air-to-ground radar tasked to report where and how enemy traffic was moving) detected the large number of vehicles massing in Kuwait City.[4] There are conflicting reports, even within the military sources, here. One Defense Department document claims that there were up to 200 tanks in this convoy.[5] This is inconsistent with post-battle inventories, which show few military vehicles in the convoy. Although we are not certain, the best explanations of this Defense Department report are either that it confused the Mutlah Ridge convoy with another farther north or that the Arab forces, which were allowed on the highway first to perform

Islamic burials of the dead, took advantage of the opportunity to "recover" as much military hardware as possible for their own use. The latter is a quite real possibility in the case of the Soviet built tanks, in particular, since they would have provided spare parts for the same kinds of tanks in the inventories of most Arab armies.

In a ground survey conducted several days after the attack, the Department of Defense confirmed that only 28 of the vehicles destroyed or left abandoned in the convoy were military.[6] The character of the vehicles is ultimately of less moral importance than the question of the *identity of the convoy drivers and passengers and the nature of the convoy's activities.*

The tactical thinking of American military commanders was heavily influenced by close study of the eight-year war with Iran that Iraq fought before its invasion of Kuwait. Iraqi tactics during the Iran-Iraq war were a source of considerable concern regarding this convoy from Kuwait city, and focused the minds of the planners on two major issues. First, the Iraqis demonstrated their will to use their superiority in armor wherever the terrain allowed for it. The Iraqi army favored a "defense-in-depth" strategy. Such a strategy required two to three layers of front line troops. These troops were used both to slow the on-coming Iranian offensive, and as "intelligence fodder" to announce what avenues of attack the Iranians were attempting to exploit. If the Iranian offensive was slowed or stalled by these forces, the rear Iraqi echelons (consisting of the highly mobile Republican Guard armor divisions) quickly crushed the offensive. The success of this type of tactic depends on both the initiative and the ability to maneuver—precisely what a retreating convoy would be attempting to gain.

The second major lesson, and the one that concerned the American commanders most, was the Iraqis' use of chemical munitions. If the situation did not allow the use of rear echelon armored divisions, then the Iraqis on several occasions retreated as quickly as possible in order to leave an open no-man's land between their forces and those of the Iranians. This area was then saturated with artillery shells containing chemical and nerve agents. The results were devastating and militarily effective. In fact, this type of attack accounted for the overwhelming success of the Iraqis in securing Iranian territory in 1988 and forcing a cease-fire.

In light of this history, planners in Riyadh considered it very possible that the Iraqi convoy at Mutlah Ridge was not running *from the Coalition forces but instead from an artillery attack by Iraqi chemical and biological weapons* they had every reason to expect would soon be launched.[7]

That night, after the JSTARS confirmation of the resistance reports, General Schwarzkopf decided not to attack the convoy right away because it was still in the city. This decision reduced the collateral damage to Kuwaiti citizens. In fact, Kuwait City was off-limits to any aerial bombardment. For obvious reasons,

---

*"The Road to Basra" is reprinted with permission from the *Journal of the Society of Christian Ethics.* pp. 213–228.

Coalition planners were politically sensitive to postwar criticisms of having "destroyed the city in order to save it."

The convoy managed to leave the city that night, move through the small town of Al-Jahra, and in the early morning hours of the 26th (still dark) was headed north to Basra near a ridge line known as Mutlah Ridge.

Now that the convoy was out in the open, the problem was no longer *whether* to attack the convoy, but *how to attack it.* The tactical difficulty was that it was dark and the weather was bad. The visibility was low and the ceiling had dropped below eight-thousand feet. Most attack aircraft had been ordered to stay above that altitude to avoid Iraqi antiaircraft fire. There was also the so-called "petroleum overcast" resulting from Iraqis setting alight Kuwaiti oilwells. These conditions required the use of the F-15E Strike Eagles. These night and all-weather air-to-ground fighters were ordered to hit the front and back of the convoy at certain choke points along the highway, immobilizing it along the road. They accomplished this mission quite successfully, and the convoy was brought to a halt by bottlenecking it at Mutlah Ridge.

The next morning, in daylight and with better weather, the "kill box" method of air interdiction was employed to funnel attacking aircraft into the Mutlah Ridge area. In this method, latitude and longitude demarcations were given that designated a thirty-mile by thirty-mile area. In this region were large numbers of mobile targets and aircraft were authorized to acquire targets and fire at will. Safe separation of aircraft was maintained by airborne command posts, which gave individual aircraft time on target commands and monitored the numbers and types of aircraft in the box at any given moment. Throughout the rest of the day, a large number of Air Force, Marine, and Navy aircraft of different types attacked and strafed this now "target-rich environment" of two to four miles length. Many pilots described the result as a "feeding frenzy" or a "turkey shoot."

There were some points of contention and bureaucratic compromise between the Air Force and the Navy concerning the command and control of aircraft. One such compromise was the Navy aircraft were not controlled by the much touted ATO from Central Command (CENTCOM) in Riyadh.[8] This central planning of the air campaign had worked well for the first three days of the air war and had maintained a single authority for targeting. According to the official Gulf War Air Power Survey, however, by this point in the war, the ATO had become a mere general outline of attacks desired, with the details filled in by local commanders.

Because of this general immunity of the Navy from the daily ATO, lines of command and control were drawn up to separate Navy and Air Force areas of responsibility for air operations. In other words, in order to insure that there were no collisions or "friendly fire" episodes between Air Force and Navy aircraft, they were given discrete geographical areas within which to operate, surrounded by buffers that, de facto, were immune to attack. The Mutlah Ridge kill box happened to be drawn right on top of this Air Force-Navy "demarcation line." (This arrangement inadvertently provided a corridor through which, on an earlier day, several Iraqi aircraft had managed to fly unhindered into Iran.) The effect of this quirk of command demarcation in the immediate vicinity of Mutlah Ridge was predictable—there was no real command and control of aircraft in and around the box. Reports and interviews with provisional wing commanders verified that several pilots failed to adhere to command and control guidelines in an effort to participate in the "turkey shoot."[9] So eager were they to join the battle that they violated standard procedures of checking in with Forward Air Controllers before entering the kill box.

Another significant aspect of this attack involved the actions of those being attacked. Several reports indicated that the Iraqis were waving white flags along the highway. But this fact must be coupled with an event that occurred two days earlier. An Iraqi unit in southern Kuwait had used the white flag as an illegitimate ruse to expose a Saudi regiment during the initial stages of the Coalition attack.[10] Iraqi forces had engaged in other such illegal tactics such as parking combat aircraft by mosques and archeological sites. Such acts of illegal perfidy by the Iraqis posed tough challenges on the commanders in the field.

As a result of these previous episodes, as well as the immediate circumstances of this attack, white flags on the highway were generally ignored (although some pilots did express reservations about this to their commanders). Even though some white flags were present, civilian vehicles in the same convoy were directing anti-aircraft fire at the attacking aircraft. Several pilots reported that the anti-aircraft fire was apparently coordinated with respect to the variable cloud ceilings throughout the day. This suggests that elements of the convoy retained at least some capability to communicate with each other and that elements, at least were still under effective command and control—*i.e.*, that they were still organized military units. In other words, despite the initial appearance of an enemy force withdrawing in chaos, the convoy, according to pilots, continued to show signs of organized retreat and command and control.

The attacks were aimed specifically at the vehicles on the highway. The goal of the air campaign throughout the theater of operations at this time was the attrition of Iraqi military hardware. This fit well with the stated U.S. national security objective of restoring security and stability to the Gulf region, which translated operationally into the destruction of the Republican Guard units of the Iraqi army.

The convoy in the area around Mutlah Ridge was the only one of several convoys throughout the Kuwaiti theater area that included a large number of civilian vehicles. It was also the only convoy that received media attention because the others were too far north inside of Iraq for press pool coverage. These convoy attacks continued into the next day (the 27th) at several other locations throughout northern Kuwait and southeastern Iraq, primarily against military vehicles, and were terminated only with the cease-fire at midnight.

2. Moral Questions Raised by Basra Road

In this section we will focus on three moral themes: 1) noncombatant immunity and the question of surrender, 2) military necessity and proportionality, and 3) observations regarding the psychology of combat and the possibilities of right intent in combatants.

First, regarding noncombatant immunity and surrender. As we saw above, popular concern that large numbers of civilian hostages were in the Basra convoy was unwarranted. Convoy participants were almost exclusively Iraqi soldiers and un-uniformed paramilitary Iraqis—and were reasonably believed to be so at the time. Although not morally decisive, they were, in fact many of the perpetrators of the worst horrors of the occupation of Kuwait City. Many military officers interviewed did appeal to the language of reprisals as justification for this attack.[11] We are not suggesting that their destruction on the road to Basra was a kind of morally justified "rough justice" for war crimes. But these facts show, at least that the objects of the attack were not innocents, either in the technical legal sense or in the more general moral sense.

There is, therefore, no question that the Iraqis on the road to Basra were not *hors de combat*. Although many in the press failed to make this distinction, it is crucial to note that they were and remained combatants. Mere armed retreat does not and should not be constructed as tantamount to surrender. Participation in such a retreat does not entitle one to any of the rights of immunity from attack granted to civilians or to surrendered military personnel. Whether we view this convoy as retreating or withdrawing, given the stated military objectives of this campaign and the imminence of the chemical threat retreat and withdrawal were synonymous and in all likelihood will continue to be viewed as such in future conflicts. Prior to any surrender or cease-fire agreement both are military maneuvers, and therefore subject to legitimate attack.

There is, however, a legitimate and important question about surrender. As we indicated above, some Iraqi troops apparently did display tokens of surrender and, in the normal case, such tokens should be accepted. But in the case at hand, the failure of the American air forces to accept these tokens seems warranted. The previous perfidious use of white flags and other indicia of surrender, the fact that flags were interspersed with elements of the convoy still engaged in coordinated hostile fire against American aircraft, and the lack of ground or even helicopter-borne troops in the area to accept the surrender and provide benevolent quarantine all justify dismissing these displays.

Perfidious use of the white flag is governed explicitly by the Annex to the Hague Conventions, summarized in the United States Army's *The Law of Land Warfare,* as follows: "Flags of truce must not be used surreptitiously to obtain military information or to obtain time to effect a retreat or secure reinforcements or to feign a surrender in order to surprise an enemy."[12] Perfidious acts are prohibited for a variety of reasons. Such acts reduce the mutual respect for the laws of war and the humanitarian principles they attempt to express. Such acts also promote the unnecessary escalation of the conflict and impede the restoration of peace. Although there does not appear to have been any centrally directed Iraqi policy to carry out acts of perfidy, actions during the Iran-Iraq conflict and several isolated incidents during Desert Storm did, in fact color the Coalition's perceptions of Iraqi attempts to offer surrender. Good faith efforts are required, insofar as they do not involve unreasonable risks to one's own troops, to determine the legitimacy of each and every white flag. In this case, it was readily apparent that surrender was not being offered on the part of entire units and that the white flags in question were the result of uncoordinated actions of individuals.[13] In light of these facts, we do not judge there to have been any moral requirement that the attack be ended because of these flags.

There is a larger in-principle question here, and one that deserves further thought and elaboration in the laws of war. There are actually no well understood conditions with respect to the concept of surrender. In fact, in the field the opposite is the case and presents yet another challenge to individual commanders and the just war tradition alike. An enemy who wishes to surrender must manifest an unconditional and unambiguous intent to surrender by way of customary indicia. Traditionally these include laying down of arms and no longer demonstrating a willingness to resist.

Unfortunately, there is no recognized universal procedure for conveying this message. One might expect and hope for a clearer description of the means of conveying this message. Yet even the current U.S. Army field manual on the law of land warfare does not delineate acceptable methods of indicating an intention to surrender. The onus at present falls on the would-be prisoner to communicate the will to surrender unambiguously.[14]

Also, even if we leave aside the perfidy question, there is a difficulty with determining whether the white flag represents the desire of the individual soldier to surrender, or that of the entire unit. The U.S. Army Field Manual, *The Law of Land Warfare,* includes the following instruction: "[The] white flag, when used by troops, indicates a desire to communicate with the enemy. The hoisting of a white flag has no other signification in international law. . . . If hoisted in action by an individual soldier or a small party, it may signify merely the surrender of that soldier or party. It is essential, therefore, to determine with reasonable certainty that the flag is shown by actual authority of the enemy commander before basing important action on that assumption. The enemy is not required to cease firing when a white flag is raised."[15] This regulation incorporates an important moral point. There is a tradeoff between allowing for the return of combatants to non-combatant status, on the one hand, and the practical constraint that although individuals may, in the face of attack, be quite prepared to quit their combatant status, surrender is normally the action of military units. Certainly, when it is practical to allow individual surrender, individuals should be granted such rights. And if, indeed, pilots deliberately targeted individual soldiers who manifested an apparent intent to surrender as *individuals* (something which we do not, in fact, know occurred) then a moral and, at least arguably, legal violation occurred. But individuals flying a white flag in the midst of an organized and armed military unit engaged in hostilities with aircraft are rarely going to find themselves in a position for surrender, nor is it practically realistic or morally requisite that the laws of war attempt to incorporate such a possibility.

On the other hand, had the facts been slightly different and entire units or even the whole column wished to surrender, there is no provision in current law or regulation that clearly indicates applicable and unambiguous indicia for such an air-to-ground engagement. Nor is it clear in such engagements what the practical implications of surrender would be, since obviously air forces can neither gain practical control over ground forces nor provide them with the benevolent quarantine required by the laws governing surrender. Hence, law and moral thought are at a conceptual limit here, and further thought needs to be given about the moral meaning of conflict and surrender during such engagements.

The second set of considerations concerns military necessity and proportionality. Was it *necessary* in military terms, to attack this convoy, or could it have been allowed safely to withdraw from Kuwait (thereby fulfilling the announced goal of the war)?

In light of the historical patterns of Iraqi tactics, and the fact that these units included some of the most elite of the Iraqi army, the attack seems well justified. As we noted above, the Coalition air commanders had every reason to fear that this withdrawal was the precursor to renewed attack or even to artillery attack using gas or biological warheads (warheads which we now know were in fact deployed and available for use with Iraqi troops).[16] General Schwarzkopf's intention was to keep such pressure on troops engaged in armed retreat that they would not be given the opportunity to regroup for attack or to set up artillery emplacements necessary to execute this tactic. Therefore, attacking the column seems well warranted indeed, and militarily desirable and necessary.

But even if necessary, was it proportionate? Probably not, at least not in the full scope of the attack. Certainly bottling up the column was warranted. The intention to destroy the armor and artillery in the column, and perhaps even the other

vehicles, seems likewise warranted. But there seems little question that gratuitous destruction was wreaked upon individual soldiers and groups far off the road and well away from the vehicles and weapons. If reports of the use of cluster bombs against soldiers on foot are true, there seems to be little justification indeed.[17]

On the other hand, determinations of proportionality require specification of both sides of the balance. In this case, much depends on what one thinks is the military and political goal of this attack. The real goal and hope in the minds of the planners went considerably further than the destruction of this unit. Indeed, the hope was the long term crippling of the Iraqi Republican Guard, that is, rendering it incapable of inflicting further damage on the Shiites of the south and the Kurds of the north. But even this intention was in careful balance with the recognition that Iraq should not be left defenseless in the face of Iran at the end of the war.[18]

The third and last topic we wish to explore here is, we think, too rarely seriously entertained by academic discussions of military ethics. This is the psychological effect of modern combat on the soldiers and airmen who fight the battles. From St. Augustine's letter to Boniface to the present day, the Christian just war tradition has always had an emphasis on the proper mental and intentional state of the warrior. Boniface, Augustine counseled, was to go to war "mournfully"—without hate or rancor, letting "necessity" and not his will do the killing.[19]

Similarly, much of the "moral armor" of the military professional consists in the belief that the destruction they bring on others is not personally or emotionally motivated, but is instead simply an instance of professionalism in conduct.

Yet much in the human dimension of the Basra Road engagement raises questions about the limits of human ability to retain such attitudes in the heat of battle. Major Hamann's interviews with many veterans of Basra Road reveal a fairly wide range of intense emotional reactions to this situation.

On the one hand, there was clearly a kind of overwhelming excitement in the minds of many pilots. Some disregarded even considerations of personal safety, neglecting to check in with and receive clearance from forward air controllers, which suggests eagerness to join in the "turkey shoot" some pilots themselves called a "feeding frenzy." Perhaps precisely because these were units directly from Kuwait City widely reported to have committed outrages against the civilians of Kuwait, the motive of revenge seems to have joined with the technical thrill of videogame-like opportunities to fire on multiple targets at will. On the other hand, some pilots felt revulsion at what they were doing and requested permission *not* to return to the scene of battle following refueling and rearming.

Besides the previously mentioned reservations about the Iraqi attempts to surrender and admissions that some pilots disregarded command and control in this engagement (aberrations in the context of the entire air campaign), the attacks on the roads to Basra revealed some startling yet persistent behaviors that can be classified into three general areas.

First some pilots delighted in the amount of destruction they could wreak on the convoy. They expressed an odd sense of pleasure in shooting a large number of live targets after weeks of destroying only hardened stationary targets. This sometimes resulted in the expenditure of a large number of antitank rounds into civilian vehicles and regret only at having wasted extra rounds of ammunition.[20] But the general delight in destruction is also a well documented psychological phenomenon of war. To quote a classic passage from Glenn Gray's *The Warriors*:

> Men who have lived in the zone of combat long enough to be veterans are sometimes possessed by a fury that makes them capable of anything. Blinded by the rage to destroy and supremely careless of the consequences . . . it is as if they are seized by a demon, or are no longer in control of themselves. From the Homeric account of the sacking of Troy to the conquest of Dienbienphu, Western literature is filled with descriptions of soldiers as berserkers and mad destroyers.[21]

This phenomenon of battle is a constant, a constant which the technological evolution of modern weapons probably will never eliminate.

Second, several pilots expressed a certain satisfaction in demonstrating the full capabilities of their aircraft on the convoys retreating into Basra. Pilots have always developed an attachment for their flying machines, even to the point where the machine seems to take on a personality of its own. This devotion varies from one type of aircraft to another, and frequently evolves into a competition amongst pilots of different weapons systems. The attacks on the roads to Basra are an illustration of this competition. Fast fighter pilots (F-15 and F-16 pilots) tend to look with disdain on the close air support mission and the aircraft that fly that mission.

The American Air Force purchased several hundred A-10 Thunderbolt II close air support aircraft in the 1970s. The pilots of those aircraft affectionately called it the "warthog" because of its ungainly appearance. Despite its appearance, the A-10 has proven to be a highly effective weapon on the modern battlefield. In fact, this aircraft was the first to be literally built around the gun it was designed to employ. The gun, a Gatling-type, which is almost twenty feet long and weighs over two tons, can fire its 30mm rounds at the incredible rate of seventy per second. It can loiter over a target area for up to one hour and is capable of destroying a significant number of tanks, armored personnel carriers, and infantry.[22]

Before Desert Storm and throughout the air campaign preceding the ground war, the A-10 had been assigned several secondary, low-priority, and low-visibility missions. While the fighters were either striking downtown Baghdad or searching for the glorious and elusive air-to-air dogfight, the results of which were guaranteed to be replayed on CNN, the A-10 was tasked for missions (search and rescue, SCUD hunts, and even battle damage reconnaissance) that it was not designed to do and its pilots were never trained to fly. Although the aircraft and crews performed those missions with distinction, the retreating convoys on the road to Basra provided them a real opportunity to demonstrate to the Air Force chain of command (primarily consisting of fast fighter pilots) the capabilities of A-10 aircraft. Several pilots boasted to Major Hamann that it was gratifying after so many years of frustration to show those idiots in command post "how much killing an A-10 could really do!"[23] A certain amount of enthusiasm is sometimes warranted and even desirable in terms of the motivation needed to face combat. But the indulgent exhibition of machines—machines with unprecedented lethality—comes precariously close to excessive violence and threatens to ignore the just war tradition's plea for restraint and virtuous conduct on the part of soldiers, even in war.

The mystique regarding soldiers' attachment to their weapons has received very little attention in the study of war. "The fear engendered by these weapons—a backdrop of unimaginable horror lying just beneath the surface—has led to a sort of paralysis of inquiry. . . . By virtue of an odd sort of consensus of opposites, weapons are generally perceived as matter-of-fact objects, mechanisms with little more symbolic and cultural significance than a pair of pliers."[24] But to understand weapons in this way is to misunderstand them.

Of Arms and Men by Robert O'Connell, a military intelligence officer and historian, explores this complex relationship between soldiers and their weapons. In his review of military history from the perspective of the evolution of weapons, O'Connell distinguishes between predatory and intraspecific aggression.

Intraspecific conflicts are characterized by ritual and ceremonial restraint, he claims. Weapons are symmetrical in size and lethality and are employed with much posturing. The virtues of the military hero in this type of conflict are coupled with an aesthetic valuing of the weapons, almost to the point where actual killing becomes a secondary objective.

In predatory conflict killing is no longer an art form. Instead, it is a mechanical process governed by an objective scientific pattern of thinking. The enemy is hunted down with casual ruthlessness, is shown no sympathy, and is not the object of feelings of shared humanity. It is here that we find language that dehumanizes the enemy.

A third observation regarding the psychology of combat was this: in Major Hamann's interviews, all of the pilots who expressed both their love of their machines and a delight in destruction also referred to the Iraqis in animal or subhuman terms. John Keegan, a noted military historian, has commented that the impersonalization of battle is one of the indices of divergence between the facts of everyday morality and battlefield morality.[25] Reflexive comments by participants on the roads into Basra are a manifestation of this index. Several pilots even went as far as to describe the enemy as a creature of indeterminate qualities—an eerie coping strategy reminiscent of the testimonials of veterans of World War II and Vietnam.[26]

There are several possible explanations for this particular psychological mechanism for coping with combat stress. All of these can be generalized as attempts to perceive as simple what is in reality very complex and, therefore, frustrating to the individual decision maker. A compartmentalized, prejudiced environment is much easier to deal with. "It must be counted as one of the particular cruelties of modern warfare that, by inducing even in the fit and willing soldier a sense of his unimportance it encouraged his treating the lives of disarmed and demoralized opponents as equally unimportant."[27] This observation brings us back to one of the most obscure of all battlefield transactions—how soldiers get their offer of surrender communicated and accepted. Insofar as these psychological dynamics are at play, clearly this too complicates that transaction.[28]

Religious and cultural differences also contributed to the common language for this dehumanization. What they were destroying were not individual persons whose plight one could sympathize with, but "camel jockeys," "rag heads," and other labels of segregation. Clearly O'Connell is right when he notes that aversion to weapons and to tactics that are deemed less than honorable is sharply reduced when the enemy is seen as fundamentally alien.

ENDNOTES

1. Hamann, interviews with Lt Col Charles D. Robertson (August 1991) and Col Christopher Christen. (May 1992) USAF CENTAF Intelligence.

2. This fact has been confirmed by the National Security Agency and Colonel Christen, the Air Force's chief intelligence officer who was in the air planning office next to Gen. Charles Homer when the information arrived.

3. Atkinson, Crusade, 438–439. Also, Desert Victory, 402.

4. James P. Coyne, Airpower in the Gulf (Arlington, VA: Aerospace Education Foundation, 1992), 169.

5. Department of Defense, Conduct of the Persian Gulf War (Washington, DC: Office of the Secretary of Defense, April 1992), 631.

6. Keaney and Cohen, GWAPS, 113.

7. During one of his Riyadh televised press briefings, when asked about the shift of emphasis of the air campaign from strategic targets to battlefield preparation, General Schwarzkopf expressed concern about the chemical threat by referring to it as "the nightmare scenario." The worst-case scenario was a principal war planning assumption and was integral to the working definition of military necessity. Reprisal chemical attacks were even considered as optional strike packages as early as August 1990. The credibility of such a threat was heightened by Saddam Hussein's public warnings. See Lawrence Freedman and Efraim Karsh, The Gulf Conflict 1990–1991:

Diplomacy and War in the New World Order (Princeton: Princeton University Press, 1993), 363. Paul Christopher, in his recent book, The Ethics of War and Peace, correctly identified the mystique that chemical weapons possess: "It is certain that chemical weapons presently cause special psychological effects that conventional weapons do not. In conversations with numerous American soldiers prior to their deployment to Saudi Arabia, I found them to be universally obsessed with the possibility of facing chemical weapons." (Englewood Cliffs, NJ: Prentice Hall, 1994), 211.

8. Keaney and Cohen, GWAPS, 153-54; Friedman, Desert Victory 173-179; Freedman and Karsh, The Gulf Conflict, 318.

9. Major Hamann, telephone interview with Lt Col Robert E. Duncan, USAF, 3 Feb 1993. Lieutenant Colonel Duncan was Director of Combat Plans, Tactical Air Control Center, Riyadh, Saudi Arabia, during the Gulf War.

10. Department of Defense, Conduct of the Persian Gulf War, 621.

11. These same officers were unaware of the additional legal requirements that various American service pamphlets specify in detail (such as the requirement that reprisals be announced to the enemy and that such actions must be authorized by national command authorities at the highest political level). Both of these conditions for a reprisal were absent here. In the absence of such constraints, this reasoning amounts to no more than an attitude of revenge or punishment which is prohibited by the laws of war.

12. Department of the Army, Field Manual 27-10, The Law of Land Warfare (Washington, D.C.: Office of the Secretary of the Army, 1956), par. 53.

13. For a full discussion of this question in another episode of the Gulf War, see Horace B. Robertson, Jr., Rear Admiral, Retired, "The Obligation to Accept Surrender," Naval War College Review 46, no. 2, sequence 342 (Spring 1993): 3-115. In this case, white flags were displayed by individuals on oil platforms. These platforms were subsequently attacked and Iraqi soldiers killed. An investigation was conducted which determined that there was no obligation to accept these "surrenders" on grounds that it was not clear these individuals had control or command over the units on the platforms. But it was additionally found that the display of flags should have been reported to higher headquarters, in case additional information was available that would have altered the perceived situation.

14. See Robertson, "The Obligation to Accept Surrender," 3–115.

15. Department of the Army, Law of Land Warfare, par 458.

16. Friedman, Desert Victory, 233.

17. "Rumors of gratuitous violence posted several years after the fact are admittedly difficult to discern as either evidence of criminal fact or misplaced accolades. But certainly boasts of firing antitank missiles at infantry troops coupled with the official disclosure that 57% of the pilots assigned to the Tactical Air Command used medical stimulants and sedatives during the war is disconcerting." Keaney and Cohen, GWAPS, 178.

18. The true objectives of the war remain contentious. The UN Resolutions never explicitly mentioned the destruction of the Iraqi Republican Guards. But by the time those resolutions were translated into specific military goals by way of the American national security objectives, destroying these divisions in the name of "regional stability" became a clear targeting priority. "There was no doubting the ultimate objective of forcing Iraq to leave Kuwait but there were questions with regard to the relationship of particular targets to this objective. As so often in the past an opportunity to mount a strategic

air campaign was seen by the USAF as a key test of airpower doctrine. . . . The campaign had objectives independent of the expulsion of Iraqi forces from Kuwait: reducing the long-term ability to exert a regional military influence and weakening the regime's hold on power" (Friedman, Desert Victory, 314). As a result of the failure to close off the escape routes to Basra and the decision by President Bush to halt the air attacks on the retreating Iraqi convoys, it has been estimated that over four Republican Guard divisions successfully retreated into Iraq. These seventy or eighty thousand troops took with them roughly eight hundred tanks, fourteen hundred armored personnel carriers, and hundreds of artillery pieces (Atkinson, Crusade, 476). "If any confirmation of the Guard's role was needed, it was provided in the uprising in Basra immediately after the war: the Republican Guard fought the Iraqi Army. It can even be argued that the course of the war strengthened Saddam's hand, in that the regular Iraqi Army was badly punished, whereas the Guard divisions were partly saved by the timing (which some would now call premature) of the cease-fire. Thus the immediate postwar balance between the strength of the Guard and the strength of the regular army was actually tilted in the direction of the Guard." (Friedman, Desert Victory, 22). These issues invite investigation of the ad bellum issues of the war and of the connection between the announced and actual war aims and the strategies and tactics employed in bello. But such discussions lie beyond the scope of this essay.

19. Augustine, "Letter to Count Boniface," in War and Christian Ethics (Grand Rapids, MI: Baker Book House, 1975), 61–63. "Of the multiple determinants of combat motivation, one of the least discussed and studied is hate. . . . Likewise, a review of the social-psychological literature on combat behavior found very little written about hate, especially in comparison to other aspects of combat behavior, such as fear and stress." John A. Balland and Alecia J. McDowell. "Hate and Combat Behavior" Armed Forces & Society, 17, no. 2 (Winter 1991): 229–41.

20. Two additional points can be made here. First, some military historians have noted that easy killing does seem to generate in humans symptoms of vindictiveness and even pleasure. See John Keegan, The Face of Battle (New York: Vintage Books, 1976), 278. Second, some psychologists have discovered evidence that shows quite clearly that the farther the initiator of aggression is from the outcome or results of his acts, the more aggressively he will act. See Ben Shalit, The Psychology of Conflict and Combat (New York: Praeger Publishers, 1988), 51. Modern warfare (especially air warfare) can be characterized as the ability to inflict severe damage at great distances. Distort that distance further with the layers of technology inherent in modern weapon systems, and the alienation can actually escalate the violent nature of such acts. With enough distance and time, crimes become misdemeanors.

21. Glenn Gray, The Warriors (New York: Harper and Row, 1959), 51.

22. The A-10 was responsible for over half of the reported and confirmed battle damage assessments in the Kuwaiti theater, even though it flew only thirty percent of the total number of Coalition sorties. See William L. Smallwood, Warthog (New York: Brassey's US, 1993), 206. Smallwood also reports that according to the debriefings of Iraqi prisoners of war, the A-10 was the single most recognizable and feared aircraft by the Iraqis (203).

23. Hamann, personal interviews with pilots, summer 1992.

24. Robert L. O'Connell, Of Arms and Men (New York: Oxford University Press, 1989), 5.

25. Keegan, Face of Battle, 320.

26. Whereas American veterans often times regarded the Japanese as vermin or insects, and the Viet Cong as the beast in the jungle, Major Hamann was taken aback when veterans of the Gulf War made reference to "the monster." Here individual enemy soldiers are collectively viewed not as a corruption or privation of good, but as an embodiment of evil. Whatever the motivations, the retreating Iraqis actually acquired the status in their minds of evil incarnate. Facing this "monster" was part of a rite of passage—a sort of personal trial that separated those who had tasted the horrors of battle and survived from those who had not. This vocabulary is hauntingly familiar to the well-known hero myth in which overcoming the monster is a condition of elevation.

27. Keegan, Face of Battle, 322.

28. Although not as extensive as some other officers, Major Hamann's personal training and experience with military war games and simulations has highlighted a shared characteristic: the lack of clarity on how to end a conflict. For example, most NATO or European simulations (which were relied upon in the initial planning of the Gulf War air campaign) either end with escalation to tactical nuclear weapons, or are left open-ended for civilian national command authorities to resolve.

INDEX